Dynamic programming is a method of solving multi-stage problems in which decisions at one stage become the conditions governing the succeeding stages. It can be applied to the management of water reservoirs, allowing them to be operated more efficiently.

This is one of the few books dedicated solely to dynamic programming techniques used in reservoir management. It presents the applicability of these techniques and their limits in the operational analysis of reservoir systems. In addition to providing optimal reservoir operation models that take into account water quantity, the book also examines models that consider water quality. The dynamic programming models presented in this book have been applied to reservoir systems all over the world, helping the reader to appreciate the applicability and limits of these models. The book also includes a model for the operation of a reservoir during an emergency situation. This volume will be a valuable reference to researchers in hydrology, water resources and engineering, as well as to professionals in reservoir management.

K . D . W . N A N D A L A L is Senior Lecturer in the Department of Civil Engineering at the University of Peradeniya, Sri Lanka. His research interests include water resources systems analysis and reservoir water quality modeling.

J A N O S J . B O G A R D I is Director of the United Nations University Institute for Environmental and Human Security. He was co-editor of *Risk, Reliability, Uncertainty, and Robustness of Water Resource Systems* (Cambridge University Press 2002).

INTERNATIONAL HYDROLOGY SERIES

The **International Hydrological Programme** (IHP) was established by the United Nations Educational, Scientific and Cultural Organization (UNESCO) in 1975 as the successor to the International Hydrological Decade. The long-term goal of the IHP is to advance our understanding of processes occurring in the water cycle and to integrate this knowledge into water resources management. The IHP is the only UN science and educational programme in the field of water resources, and one of its outputs has been a steady stream of technical and information documents aimed at water specialists and decision-makers.

The **International Hydrology Series** has been developed by the IHP in collaboration with Cambridge University Press as a major collection of research monographs, synthesis volumes and graduate texts on the subject of water. Authoritative and international in scope, the various books within the series all contribute to the aims of the IHP in improving scientific and technical knowledge of fresh-water processes, in providing research know-how and in stimulating the responsible management of water resources.

Dynamic Programming Based Operation of Reservoirs

Applicability and Limits

K. D. W. Nandalal

University of Peradeniya, Sri Lanka

Janos J. Bogardi

United Nations University, Bonn, Germany

CAMBRIDGE
UNIVERSITY PRESS

CAMBRIDGE UNIVERSITY PRESS
Cambridge, New York, Melbourne, Madrid, Cape Town, Singapore, São Paulo

Cambridge University Press
The Edinburgh Building, Cambridge CB2 8RU, UK

Published in the United States of America by Cambridge University Press, New York

www.cambridge.org
Information on this title: www.cambridge.org/9780521874083

First published 2007

Printed in the United Kingdom at the University Press, Cambridge

A catalog record for this publication is available from the British Library

ISBN 978-0-521-87408-3 hardback

Contents

Figures

Tables

Preface

The second half of the twentieth century can clearly be identified as an epoch having a strong, lasting imprint on our paradigms and methods of resource use and management. Ideas, compassions, and concepts which dominate our thinking and debates have emerged and evolved during the last four or five decennia. Nothing manifests this better than the so-called Brundtland Report (WCED, 1987). Ever since its publication, the term and concept of sustainable development cannot be missed in any declaration or framework issued or developed in seeking better conditions for humans and the environment alike. The recent millennium was a welcome opportunity to summarize this process and endorse principles and set new objectives. As far as the ethical, political, and practical aspects of water resources management are concerned, the large intergovernmental environmental conferences like the United Nations Conference on Environment and Development (UN, 1992) and the World Summit on Sustainable Development (WSSD, 2002) can be mentioned along with the formulation of UN Millennium Development Goals (MDGs, 2000) and the Millennium Ecosystem Assessment (2005). Beyond these general conferences and assessments, where water took a substantial part of the agenda, the world water fora (Marrakech, 1997; The Hague, 2000; Kyoto, 2003; Mexico City, 2006) and the Bonn Conference on Freshwater 2001 provided the broadest platforms for stakeholder dialogue involving ministerial, NGO, scientific, professional, and other interest groups, and indigenous people participation. The impacts of these conferences were analyzed by, among others, Bogardi and Szöllösi-Nagy (2004).

Besides these events, the World Water Vision (Cosgrove and Rijsberman, 2000) and the first issue of the World Water Development Report (2003) can be mentioned as the key documents, summarizing the process of assessing the availability, use, and protection of this precious resource. Irrespective of considerable successes in putting water issues on the international agenda (such as the Group of 8 meeting in Evian in 2003), we are far from having secured the "breakthrough" towards achieving the water related MDGs and other global objectives.

A book like the present one, focusing on one methodological concept and its use in a particular form of water resources management, namely the application of dynamic programming (DP) in the operational analysis of reservoirs, would certainly be overcharged if not only the principles and the history of the idea of sustainable development, MDGs, environmental awareness and protection, and biodiversity, but also water supply, food and energy security, disaster mitigation or participatory processes, public–private partnerships, and other key issues of the present water debate were presented and discussed in the full context of their historic evolution. Yet these two lines of thought, the conceptual one describing our changing world views, and the more focused methodological development of management techniques – in this case the application of DP – are closely intertwined. Resource limitations and increasing demand pose the question of human and ecosystem survivability and reveal the urgent need for better tools and methods to match resources and demand at a certain point in space and time on the practical governance (management) scale.

Even if we concentrate only on the subject (and the inherent self-limitations) of this book, a 50-year-long saga unfolds. While storing water is certainly among the very first actions of human civilization (aptly proven by remnants of dams from antiquity) the 1950s (and the following three decades) experienced the strongest boom ever in dam building. Almost three-quarters of the dams of the worldwide total of approximately 40 000 were built between 1950 and 1980 (Takeuchi, 2002). The storage capacity thus created in many parts of the world – while not uncontroversial in its environmental impacts and other side effects – has certainly contributed to avoiding worst-case-scenario prophecies of food shortage at global scale.

However, building dams alone could not and cannot solve the problem. Half a century ago we paid more attention to sound engineering of the structures than to efficient management of the then new facilities, or to erosion control in the upstream watersheds. Consequently, the potential of many reservoirs was not exploited to the full. Instead of refining operational rules, saving water, and saving storage space from being lost to siltation, more and more new dams were built. No wonder that, with growing environmental awareness and international eco-advocacy, the Hamletian question "to build or not to build?" was answered more and more by choosing the latter option. The creation of the World Commission on Dams (WCD), its report *Dams and Development* (2000) and the subsequent reactions of professional associations like ICOLD and ICID mark this process. In the meantime much less attention was given to the less dramatic, but nevertheless crucial question: "Do we operate our reservoirs well?" The answer to this silent question would have been and, regrettably enough, would still be no rather than yes. While the first part of this "double no," not to build new dams and not to use the existing ones to their fullest potential, could be seen as ideologically biased; the second "no" is actually unforgivable, irrespective of one's position as pro or contra dams. Improving the performance of existing reservoirs and complex reservoir systems would not only provide more water for more beneficial uses, but could also mitigate environmental impacts and significantly reduce the need for new dams. Thus a proactive approach to improve reservoir operation would ultimately ease, if not eliminate, the urgency of some "build or not to build" dilemmas.

Do we have the means to implement the necessary improvements? It is the conviction of the authors that the answer must be a resounding yes, an opinion that we believe is broadly shared by the respective scientific community.

Almost parallel to the previously described dam-building boom systems analysis, operations research (OR) techniques have emerged as new intellectual tools with which to analyze complex systems. The introduction of digital computational technology and what we today call information technology opened the door for wide-scale, practically relevant applications. As far as dynamic programming, the OR method with the biggest potential to improve reservoir operation, is concerned, the year 2007 has special significance. It marks the 50th anniversary of the pioneering paper by Bellman (1957) formulating and proving the optimality criterion of this appealing decomposition technique. This book is dedicated to observing this anniversary. Yet there is no real ground for celebration beyond commemorating a significant scientific achievement. This milestone could and should be taken as an opportunity to review why 50 years in the emerging information society, with its fast knowledge transfer mechanisms, thousands of papers, articles, lectures and conference presentations, and dozens of successful case studies, did not suffice to ensure a wide-scale breakthrough of DP based methods into real-world reservoir system operation.

The advent of desktop computational development in the 1980s and 1990s brought the opportunity for research groups to prove that DP and its derivative methods are not only exciting scientific tools, but potent techniques to be applied in improved reservoir operational management worldwide. This book confirms this peak, as most of the references originate from the last two decades.

There is an inherent and acceptable time lag between scientific discovery and development and "real-world" application. However, the students of the 1980s and 1990s are already in management positions, thus the question needs urgent attention: why have we only a handful of real practical applications like the DP based operational analysis of the reservoir system in the strategic plans of "Eau 2000" and "GEORE" of Tunisia (Bogardi *et al.*, 1994).

This book emerges from the concern of those actively involved in the development of DP based operational methods for reservoir systems. Many of the practically relevant case studies, tests of DP and stochastic dynamic programming (SDP), were carried out between 1985 and 1998 at the Asian Institute of Technology, Bangkok, Thailand, and later at the then Wageningen Agricultural University, the Netherlands, under the supervision and guidance of the second author. It is however due to the enthusiasm and dedication of the first author that this present book came into being. He not only initiated but also carried out the most overwhelming part of the work, which provides a comprehensive account of the applicability of DP based methods to derive sophisticated and yet practically relevant rules for real-world reservoir systems, operating under real-world conditions and constraints.

This work, while reflecting the entire related literature, is reliant on results published in several reports, papers, dissertations, and master theses prepared in the late 1980s and 1990s by several members of the above-mentioned research groups. The authors wish to acknowledge the implicit intellectual input and active assistance of Dr. Saisunee Budhakooncharoen, Professor Huang Wen Cheng, Dr. M. D. U. P. Kularathna, and Dr. Darko Milutin. The works of He Qing, Anne Verhoef, Dr. Bijaya Prakash Shrestha, and Dr. Dinesh Lal Shrestha are also reflected in this book. Furthermore, collaboration with Professor Ricardo Harboe, Dr. Guna Nidi Paudyal, and Professor Ashim Das Gupta as co-authors of papers and co-supervisors of some of these theses is greatly appreciated.

The aim of this book goes beyond providing the reference for our claim that DP based techniques can and should be applied for the improved operation of reservoir systems, even under conditions of changing objectives, constraints and hydro-climatic regimes as has been demonstrated recently by

Brass (Brass, 2006). We feel that, next to its contribution to bridging the gap between development and method applications, the book could be used as special reading for graduate students specializing in water resources management. In this context, this book can be seen as an extension of DP related methods and reservoir system operation supplementing the excellent textbook of Daniel P. Loucks and Eelco van Beek, *Water Resources Systems Planning and Management* (Loucks and Beek, 2005).

It is our paramount objective to contribute to the education of competent water resources managers. This book is intended to be an eye-opener for those bearing managerial responsibilities at present, and a source of inspiration and knowledge for the coming generation of water resources managers.

1 Water resources management

1.1 GENERAL

The water resource has a major influence on human activities. It is a major input in almost all sectors of human endeavor. Water serves essential biological functions and no human can survive in its complete absence. Water's contributions to human welfare include its role as a basic element of social and economic infrastructure. Also important are water's natural attributes that contribute to human aesthetic enjoyment and general psychological welfare. But water also has negative impacts on human well-being. Floods, inundations, and water-borne diseases are also associated with water.

Water has played a major role in socio-economic development due to the magnitude and widespread occurrence of its positive and negative impacts. The quality of human life is directly dependent on how well these resources are managed. Water management activities are intended to enhance the positive contributions of water or control its negative impacts.

Ancient civilizations grew up in the river valleys of the Tigris and Euphrates, Nile, Indus, Yellow River, etc., where there was plenty of water. Water management activities, particularly irrigation, played a central role in the development of these civilizations. In those days the planning and management of the water resources were primarily for single uses. The continuing growth of the human population, especially since the nineteenth century, together with rapid industrial development and rising expectations of a better life necessitated more complex and consistent water resources management. These competing demands and uncontrolled use, along with the pollution of water, have made it a scarce resource.

Water resources problems are going to be more complex worldwide in the future (Simonovic, 2000). Population growth, climate variability, regulatory requirements, project planning horizons, temporal and spatial scales, social and environmental considerations, transboundary considerations, etc., all contribute to the complexity of water resources planning and management problems. Traditional engineering has gradually been overchallenged by the multitude of claims, constraints, and opportunities. Since the Second World War, systems analysis has emerged as one of the tools for solving such complex water resources management problems (Dantzing, 1963; Hillier and Lieberman, 1990; Loucks et al., 1981).

Systems analysis can generally be defined as a group of methods developed for identifying, describing, and screening a system, its performance and behavior under different conditions and with different goals to be pursued. It provides a decision maker with a broad information base about the system and gives the opportunity of estimating the system behavior to compare several feasible alternatives. In its process, a variety of initial assumptions, objectives, constraints, and decision variables are specified and their influence on the system operation is evaluated. Hence, systems analysis techniques can be very valuable tools for solving planning and operating tasks in water resources management based on the systematic and efficient organization and analysis of relevant information.

There are a number of terms which are used synonymously with the systems approach; these include systems engineering, operations research, operations analysis, management science, cybernetics, and policy analysis. Hall and Dracup (1970) defined systems engineering as the art and science of selecting, from a large number of feasible alternatives involving substantial engineering content, that particular set of actions which will accomplish the overall objectives of the decisions makers, within the constraints of law, morality, economics, resources, political and social pressures, and laws governing physical life and other natural sciences.

Together with the determination of physical elements of a system, the operation policy of the system is equally important in finding the best performance of the system to serve its purpose. The operation policy of a water resources system can be defined on a short-term or a long-term time base. This classification implies not only the time base (e.g., hourly or daily for short-term and monthly or seasonal for long-term operation) but also the uncertainty of the system and its components. For short-term operation, uncertainty may be neglected, and all the phenomena can be considered as deterministic. However, for long-term operation the stochasticity, inherent both in a system and in its environment, must not be neglected. The complexity of a system itself, together with the uncertainty of all the phenomena involved including the goals to be achieved, raises the need for effective methods for deriving such operation policies that would provide an expected "optimal" response of the system under a number of different conditions. A variety of methods in systems analysis or operations research have been developed for analyzing water resources systems. In general, systems analysis implies two basic strategies in operational assessment: simulation and optimization approaches.

Simulation is used to analyze the effects of proposed management plans: achievement regarding system performance is evaluated based on selected sets of decisions. By definition, the simulation method does not claim that a particular combination of decisions represents the optimal one. The difficulty inherent in this approach is the large number of feasible operation plans (combinations of decisions) to be checked. If simulation alone were used, the search for the "best" solution might not only be very tedious, but also could lead to alternatives far from the optimal one.

Optimization models are used to narrow down the search for promising combinations of decision variables. Optimization eliminates all the undesirable operation plans and proposes policies which are close to the global optimal solution. However, optimization usually relies on a very simple representation of a water resources system. Therefore, optimized alternatives may be further refined by applying simulation techniques. The most frequently used optimization techniques in water resources management can be classified into three major groups: (1) linear programming (LP), (2) dynamic programming (DP), and (3) nonlinear programming (NLP). This general classification, in addition to simulation models, represents the basic methods used in planning and management of water resources systems (Yeh, 1985). An extremely large number of simulation and optimization models providing a broad range of analysis capabilities for evaluating reservoir operations have been built over the past several decades. Wurbs (1993) sorted through these numerous models and reached a better understanding of which method might be the most useful in various types of decision support situations. Since most of the water resources systems display considerable nonlinearities and sequential nature, operational assessment – especially in the case of reservoirs – is usually based on DP. The more so, since DP lends itself to a relatively easy incorporation of stochasticity (Loucks et al., 1981).

1.2 ROLE OF RESERVOIRS

According to Takeuchi (2002), there are presently nearly 40 000 large reservoirs in the world impounding approximately 6000 km^3 of water and inundating an aggregate area of 400 000 km^2. Recent surveys show that this number increases at a rate of approximately 250 new reservoirs each year. These figures clearly reflect the fact that reservoirs, irrespective of their interference in the aquatic ecosystem of the respective watercourse, have a firmly established position in our striving to harness and manage the available water resources.

The history of man-made reservoirs can be traced back to antiquity. Perhaps at the beginning the "water reservoir" was no more than a huge tank to store water during the wet season for use during the dry season. Today, with the development of civilization, reservoirs can be found all over the world. The reservoirs can serve single or multiple purposes including hydropower generation, water supply for irrigation, industrial and domestic use, flood control, improvement of water quality, recreation, wildlife conservation, and navigation. The effective use of reservoir systems has become increasingly important. Next to the exigence of the rational use of a limited resource, a better-managed reservoir may make the physical extension of the system – to add new reservoirs – unnecessary. The operation of a single reservoir for a single function does not present many analytical problems, but the same is not true when a reservoir fulfils a number of potentially conflicting objectives or where several reservoirs are operated conjunctively. Through a global review of performance of dams/reservoirs, the World Commission on Dams (2000) presented an integrated assessment of when, how, and why dams/reservoirs succeed or fail in meeting development objectives.

Reservoir construction was most intensive during the period 1950–70 in many well-developed regions where river runoff was finally almost fully regulated. Subsequently, the rates of reservoir construction have decreased considerably although they are still high in those countries with rich natural resources of river runoff. This is caused partly by the increasing role of hydropower engineering where there are liquid and solid fuel deficits. In addition, reservoirs provide the greater part of the water consumed by industry, power stations, and agriculture. They are the basis for large-scale water management systems regulating river runoff as well as protecting populated areas from floods and inundations.

1.3 OPTIMAL RESERVOIR OPERATION

Reservoirs have to be best operated to achieve maximum benefits from them. For many years the rule curves, which define ideal reservoir storage levels at each season or month, have been the essential operational tool. Reservoir operators are expected to maintain these pre-fixed water levels as closely as possible while generally trying to satisfy various water needs downstream. If the levels of reservoir storage are above the target or desired levels, the release rates are increased. Conversely, if the levels are below the targets, the release rates are decreased. Sometimes operation rules are defined to include not only storage target levels, but also various storage allocation zones, such as conservation, flood control, spill or surcharge, buffer, and inactive or dead storage zones. Those zones also may vary throughout the year and the advised release range for each zone is provided by the rules. The desired storage levels and allocation zones mentioned above are usually defined based on historical operating practice and experience. Having only these target levels for each reservoir, the reservoir operator has considerable responsibility in day-to-day operation with respect to the appropriate trade-off between storage levels and discharge deviations from ideal conditions. Hence, such an operation requires experienced operators with sound judgment. Needless to say, predetermined operation rules have proven to be quite inflexible when dealing with unexpected situations.

To counteract the inefficiency in operating a reservoir system only by the "rule curves," additional policies for operation have now been incorporated into most reservoir operation rules. These operation guidelines define precisely when conditions are not ideal (e.g., when maintenance of the ideal storage levels becomes impractical), and the decisions to be made for various combinations of hydrological and reservoir storage conditions. For some reservoir systems, this type of operation policy has already taken over the rule curves and is acting as the principal rule for reservoir operation.

Over the past several decades, increasing attention has been given to systems analysis techniques for deriving operation rules for reservoir systems. As the references reveal, the 1980s and 1990s were the most productive period in this respect. As a result, a variety of methods are now available for analyzing the operation of reservoir systems. In general, these techniques lead to models which can be classified into two categories: optimization models and simulation models. Simulation models can effectively analyze the consequences of various proposed operation rules and indicate where marginal improvements in operation policy might be made. However, the simulation technique is not very appropriate in selecting the best rule from the set of possible alternatives.

Optimization models can eliminate the clearly undesirable alternatives. Yeh (1985) reviewed the state-of-the-art of the mathematical models developed for reservoir operations. The alternatives that are found to be most promising based on optimization methods can then be further analyzed and improved using simulation techniques.

Although both optimization and simulation can be, and at times are, used independently to analyze an operational problem, they are essentially two complementary methods. In fact, optimization and simulation are used conjunctively to derive and to assess alternative operating strategies of single and multiple reservoir systems (e.g., Jacoby and Loucks, 1972; Mawer and Thorn, 1974; Gal, 1979; Karamouz and Houck, 1982, 1987; Stedinger et al., 1984; Tejada-Guibert et al., 1993; Harboe et al., 1995; Liang et al., 1996).

Linear programming (LP) and dynamic programming (DP) have been the most popular among the optimization models in deriving optimum operation rules for reservoir systems. Linear programming is concerned with solving problems in which all relations among the variables are linear, both in the constraints and in the objective function to be optimized. The fact that most of the functions encountered in problems with reservoir operation are nonlinear has been the main obstacle to the successful and practically relevant use of LP in this area. Although linearization techniques can be employed, this might not be satisfactory. The degree of the approximation required in the linearization process can seriously affect the reliability associated with this technique. However, LP has been used in optimal reservoir operation and the following are some applications: Gablinger and Loucks (1970), Roefs and Bodin (1970), Gilbert and Shane (1982), Shane and Gilbert (1982), Palmer and Holmes (1988), Randall et al. (1990), Reznicek and Simonovic (1990, 1992). Due to this favorable coincidence the authors are convinced that dynamic programming and its derivative techniques have a superior applicability to serve as the basis for the operation of real-world reservoir systems. Hence this book is dedicated to exploring this potential. More than two decades after the large-scale introduction of DP based reservoir operational studies, the time has come to review the development and to outline, supported with practical case studies, the vast field of applicability of DP based rules in reservoir system operation.

Dynamic programming, a method that breaks down a multidecision problem into a sequence of subproblems with few decisions, is ideally suited for time-sequential decision problems such as deriving operation policies for reservoirs.

1.4 CONVENTIONAL DYNAMIC PROGRAMMING

Dynamic programming is a technique used for optimizing a multistage process. It is a "solution-seeking" concept which replaces a problem of n decision variables by n subproblems having preferably one decision variable each. Such an approach allows analysts to make decisions stage-by-stage, until the final result is obtained. Hence the original problem needs to be decomposed into subproblems and each subproblem is referred to as a stage. This decomposition could be defined either in space or in time. Each stage is characterized by different system states expressed by the numerical value of selected state variable(s). Transition of the state from one stage to the next is expressed by a particular course of action (or the decision what to do), which is represented by a decision variable. Changes of the system's state influenced by the decision taken at the previous stage are described by the state transformation equation. This transition of the state is possible only if certain rules are followed: both system state and decision variable can take values within particular domains. These limits form a set of constraints which must be met at every stage during the optimization process.

The computational routine for deriving the optimal policy follows Bellman's recursive equation (Eq. 1.1), which is described diagrammatically in Figure 1.1. This can be solved by either moving forward (forward DP) or moving backward (backward DP) stage by stage.

For every state s at stage j the optimal policy is given by (subscripts denote backward computational procedure)

$$f_j^*(s_j) = \max_{x_j} \left\{ C_{S_j X_j} + f_{j+1}^*(s_{j+1}) \right\}, \qquad (1.1)$$

where

$C_{S_j X_j}$ = costs or contribution of the decision X_j given state S_j at the actual stage,

f_{j+1}^* = accumulated suboptimal costs (or contribution) for following stages $j+1, j+2, \ldots, N$,

N = total number of stages,

s_j = system state at stage j,

$s_{j+1} = t(s_j, x_j)$ = state transformation equation,

j = stage, and

x_j = decision taken at stage j.

In other words the above equation reflects Bellman's principle of optimality. Generally, the DP procedure starts by setting the objective function's value (cost or benefit) at the initial stage to zero, or any other arbitrary value. Subsequently, suboptimal policy derived at the last computational stage is

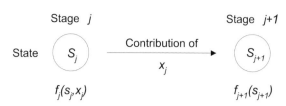

Figure 1.1 Basic structure of dynamic programming

actually the global optimum of the problem. The optimal policy can then be derived as a set of decisions, each of which is taken at a subsequent stage with respect to the corresponding suboptimal decisions derived at the preceding stage.

It is essential to point out that DP models require problem-specific formulations. This is due to differences that appear among the variety of problems that can be solved using DP: objective functions can have different forms; some problems have one and some of them can have several state variables; state transformation equations are not the same in all cases; decision variables can vary among different problems, etc.

1.5 INCREMENTAL DYNAMIC PROGRAMMING

Simultaneous derivation of operation policies for all the reservoirs in a multi-reservoir system is important, because the optimum conditions of the system cannot be investigated by considering reservoirs in isolation. In conventional DP, the state variables (reservoir storage) are normally discretized. Dense discretization is preferred over a coarse one to obtain an operation policy close to the global optimum. These two factors, simultaneous investigation of all the reservoirs of the system (state variables) and dense discretization of these state variables, exponentially increase the total number of state variables to be considered. This phenomenon is termed the "curse of dimensionality" of DP problems.

Larson (1968) introduced incremental dynamic programming (IDP), a successive approximation method, to overcome high dimensionality problems. This chapter presents the IDP technique. Several applications of the IDP technique in reservoir management are presented in subsequent chapters.

Incremental dynamic programming is one of the techniques used in alleviating the problems of excessive time and computer storage requirements. The general scheme of IDP procedure is concisely presented in Figure 1.2. IDP uses the recursive equation of DP to search for an improved trajectory starting with an assumed feasible solution, which can be visualized as a trial trajectory. The improved trajectory is then sought within the pre-specified range, defined as the "corridor."

The computation cycle is complete when the search process has converged to the optimal trajectory according to a pre-specified convergence criterion. New iteration steps are needed as long as the convergence criterion is not satisfied. In the next iteration the locally improved trajectory obtained from the previous iteration serves as the new initial trial trajectory.

The IDP procedure begins with selection of a trial trajectory. A trajectory is the sequence of admissible transformations of the state vectors throughout the entire period of consideration. It also defines the initial value of the objective function. A trajectory is feasible if it satisfies all constraints. It is optimal if it is associated with the best possible achievement of the objective criterion of the system performance.

The basic idea behind the selection of an initial trajectory is to provide, for the search process for the optimal trajectory, both a starting point and a region called the "corridor" around the trial trajectory. The initial trial trajectory should therefore be feasible since it serves as the first approximation of the optimal trajectory.

The next step of the IDP procedure after determining an initial trial trajectory is construction of a corridor around it as shown in Figure 1.3.

The corridor specifies the values of state variables to be considered at each time step in the optimization process. For a given corridor, the difference between adjacent values of a state variable is the width of corridor. In general, a corridor for a single-reservoir system consisting of three state variables is defined symmetrically around the trial trajectory of state variable S_j as follows:

$$UBC = S_j + \Delta, \tag{1.2}$$

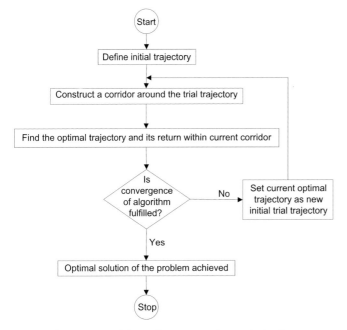

Figure 1.2 Incremental dynamic programming optimization procedure

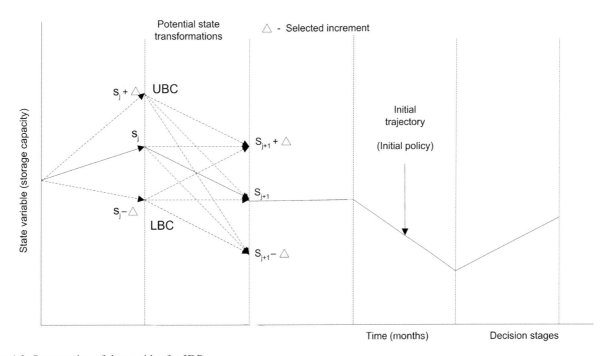

Figure 1.3 Construction of the corridor for IDP

$$LBC = S_j - \Delta, \qquad (1.3)$$

where
UBC = upper bound of corridor,
LBC = lower bound of corridor,
Δ = half corridor width, and
S_j = state variable at the beginning of stage j based on the initial policy (first trial).
However, nonsymmetrical corridors may result if any of the boundaries of the corridor exceed the feasible state space of the system.

After the construction of a corridor around the trial trajectory, an improved trajectory and the corresponding objective function value within the corridor are sought. This is done by using the recursive equation of the conventional DP algorithm restricting computations of the state transformations to prefixed values of state variables within the pre-specified corridor.

Convergence behavior of the IDP algorithm to reach the global optimum is an essential issue. Selection of the feasible initial trial trajectory is entirely an arbitrary process. But in standard practice the initial corridor width is a coarse one. This technique follows the principle of choosing the initial corridor width sufficiently large to cover a considerable range of potential storage volume for the first cycle of the IDP procedure. The corridor width is decreased progressively as the iteration proceeds (Turgeon, 1982).

In general, the larger the initial corridor width around the initial trajectory, the smaller the number of iterations required to reach the optimal solution. Use of a large corridor width in earlier iterations is to ensure that the improved trajectories for such iterations are really obtained. Moreover, since the initial trajectory for any later iteration is the improved trajectory compared to the preceding one and it is closer to optimality, smaller corridor widths can be used in later iterations to search for the optimal trajectory.

The iterative process is then continued until a convergence criterion, explained later, is fulfilled. The objective function value obtained after termination of the IDP is considered as the optimum value. To ensure that the final solution is a true optimum value, a few more sets of IDP computation runs with different initial corridor widths may be attempted and the results compared to check whether the solution obtained remains the same.

According to the IDP procedure, each iteration of search for the improved trajectory results in a trajectory which is associated with a better value of the objective function than that of the trajectory for the preceding iteration. The convergence of the IDP solution exhibits a monotonic nature. Thus, a point convergence cannot be attained unless the cycle of computation is allowed infinitely. Therefore, the convergence criteria should be defined to limit the computer time used.

The iterative process of IDP is repeated until there is no further significant improvement of objective function value. As a criterion to terminate the computation, the following expression can be applied. That is, whenever

$$\frac{(\text{OF}_i - \text{OF}_{i-1})}{(\text{OF}_{i-1})} \le 0.0001 \qquad (1.4)$$

then the computation cycle should be terminated.

Here, OF_i is the objective function's value with respect to the set of constraints for iteration, $i = 1, 2, 3, \ldots$

Instead of searching for the optimal solution over the entire state-space domain as in the classical DP, only three states of storage volume are involved in the analysis at any iteration in the case of a single reservoir. Similarly, IDP can tackle multiunit reservoir systems by taking a limited state space for every individual reservoir in the system. Thus, this technique can overcome the problem of dimensionality. Computer storage and computer time requirements can be reduced considerably.

1.6 STOCHASTIC DYNAMIC PROGRAMMING

Stochastic dynamic programming (SDP) is very common in reservoir operation. Since uncertainty is the inherent characteristic of water resources systems, it is often inadequate to opt for deterministic decision models, at both planning and operational stages.

The stochastic nature of inflows can be handled by two approaches: an implicit or an explicit approach. In the implicit approach, a time series model is used to generate a number of synthetic inflow sequences. The system is optimized for each streamflow sequence and operating rules are found by multiple regression. During the optimization the synthetic data series are considered as deterministic series. The implicit approach optimizes the system operation under a large number of streamflow sequences, at the expense of computer time.

The explicit approach considers the probability distribution of the inflows rather than specific flow sequences. This approach generates an operation policy comprising storage targets or release decisions for every possible reservoir storage and inflow state combination in each time step, rather than a mere single schedule of reservoir releases.

Future states or outcomes of any stochastic process such as rainfall and streamflow cannot be predicted with certainty. However, based on past performance, probability associated with any particular outcome may be estimated. Hydrological uncertainty of streamflows is explicitly taken into consideration in the explicit SDP models. These models incorporate discrete probability distributions in the optimization process. They describe the extent of uncertainty of future occurrences of streamflows and correlations of streamflows in time and space that may be present among streamflow time series to different reservoirs of the same water resources management system.

Assuming that the unconditional steady-state probability distributions for monthly streamflows are not changing from one year to the next, a Markov chain could be defined for streamflow. Since there are 12 months in a year there would be a lag-one Markov chain with 12 transitional probability matrices. The elements of it could be denoted as $P_{p,q}^j$, the probability of occurrence of a streamflow class q in month $(j + 1)$ given a streamflow state p in month j. In the model presented, first order (lag-one) Markov chains are used to estimate the discrete conditional (transition) probabilities that represent the stochasticity inherent in streamflows. Discrete transition probabilities are estimated for a number of representative inflow values for each month, using the available historical streamflow records.

In a DP formulation of a reservoir operational problem, time periods are often considered as stages. The stored volumes of water in reservoirs at the beginning of the time periods represent the state of the system. The decisions to be taken at each stage are the quantities of water to be released. These can be implicitly identified by specifying the storage volumes at the next stage (identifying the storage volumes at the end of the time step considered). To incorporate the markovian nature of the streamflow, it is also defined as a state variable in SDP formulations. Therefore, a SDP formulation of a reservoir operational problem will have a two-dimensional state space representing the storage volumes and inflows to the reservoirs.

Use of SDP requires discretization of state variables and representation of them by a finite number of characteristic values. Sets of characteristic (representative) storage volumes and streamflows are chosen to cover the entire range of possible storage volumes and streamflows.

The domain of inflows, which must be wide enough to represent the entire range of potential inflows, is divided into a certain number of intervals or classes. These intervals or classes could be equally spaced or of variable size. In general, averages of the inflows that fall into these intervals are chosen as discrete values to represent inflow classes. The values represent the entire interval in the subsequent computations.

Means and variances of inflows during each month can be used to check whether they are reproduced by the discretization.

If they are found to be not reproducing these statistics satisfactorily, a trial-and-error selection of the class margins and representative values may be used. Frequency diagrams can be of help in the selection procedure.

Interval $(S_{j,\min}, S_{j,\max})$ is divided into NS − 1 equally spaced storage intervals, where $S_{j,\min}$ and $S_{j,\max}$ are the minimum and maximum limits of live storage of the reservoir at the beginning of month j. NS is the number of reservoir space classes. Then the boundary values of these equally spaced intervals are used as discrete values of storage.

The backward stochastic dynamic programming algorithm (Loucks *et al.*, 1981) is used for optimizing reservoir operation. The forward algorithm has no sense in the case of SDP, as the expectation over the future states has to be considered. The SDP optimization process derives the optimum operating strategy of the reservoir from Bellman's backward recursive relationship:

$$F_j^n(S_j) = \mathop{\text{Opt}}_{X_j}\left\{ B\big(S_j, S_{j+1}, I_j\big) + \sum_q P_{p,q}^j \times F_{j+1}^{n-1}\big(S_{j+1}\big)\right\},$$

(1.5)

where
$B(S_j, S_{j+1}, I_j) =$ cost or contribution of the decision X_j given state S_j at the initial stage,
$F_{j+1}^{n-1} =$ accumulated suboptimal cost (or contribution) by optimal operation of the reservoir over the last $n − 1$ stages,
$I_j =$ inflow during period j,
$P_{p,q}^j =$ transition probabilities of inflows (defined previously),
$S_j =$ system state at stage j,
$S_{j+1} = t(S_j, X_j) =$ state transformation equation,
$j =$ stage, and
$X_j =$ decision taken at stage j.
The outline of the SDP procedure is displayed in Figure 1.4.

The SDP procedure starts by initiating the values of the objective function at the last stage (a month in the future) to zero, or any other arbitrary value, for each combination of the discrete values of the two state variables at some time step in the future. Thereafter the process continues by traversing backwards along the temporal stages (i.e., months). The optimization consists of a number of iterations, each having 12 monthly stages representing one annual cycle. Usually one iteration cycle comprises 12 stages (months) of computation but more refined temporal stages (decades, etc.) can also be envisaged. The aggregate of the objective function's expectation grows by setting its value at the beginning of each iteration (i.e., a year) to the respective accumulated value of the objective function at the end of the last stage of the previous iteration. After a few iterations, the increase in value for any state over a period of one year becomes constant

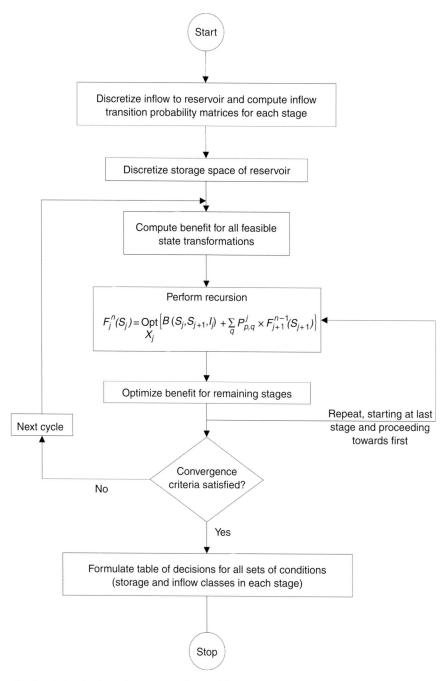

Figure 1.4 Flow diagram for the stochastic dynamic programming model

and independent of the state. This is the expected annual return from the operation of the system.

There are two criteria that determine the convergence.

(a) Stabilization of the expected annual increment of the optimum value obtained by Bellman's recursive formula (Loucks *et al.*, 1981).

During continued backward computation of the SDP algorithm, the optimum expected return for all possible initial

states will be determined for each stage (month). When the expected return for a period of one year becomes constant for all state transformations in each stage (month), the convergence criterion of constant expected annual objective achievement is satisfied.

(b) Stabilization of the operation policy (Chow *et al.*, 1975).

At each stage (month) of the SDP algorithm, an operation policy for that stage is determined. After continuing backward computation for a couple of annual cycles, a stable operation policy can be obtained. This implies that once stabilized the operation policy for a specific month will not change from year to year. When this condition is reached the convergence criterion of stabilization of the operation policy is achieved.

Operation policy designated by SDP is a set of rules specifying the storage level at the beginning of the next month for each combination of storage levels at the beginning of the current month and inflow during the current month. Due to the discrete nature of the SDP algorithm, the number of state transformations in any stage shows an exponential growth with increase of the number of state variables. A polynomial growth of the number of state transformations at each stage can be noted with increase of the number of state discretizations. This is reflected in the excessive computer time and memory requirements necessary to run a SDP model with a comparatively fine discretization of state variables.

1.7 DYNAMIC PROGRAMMING IN RESERVOIR OPERATIONS

Dynamic programming is an approach developed to solve sequential, or multistage, decision problems; hence the name "dynamic" programming. But this approach is equally applicable for decision problems where the sequential property is induced solely for computational convenience.

The DP technique is efficient in making a sequence of interrelated decisions. It is based on Bellman's principle of optimality (Bellman, 1957): "An optimal policy has the property that whatever the initial state and the initial decisions are, the remaining decisions must constitute an optimal policy with regard to the state resulting from the first decision." That implies a sequential decision process in which a problem involving several variables is broken down into a sequence of simpler problems, each having preferably a single variable. DP is very well suited to studying reservoir operational problems. Since its development, the number of applications of DP in studying reservoir operational problems has increased enormously. The DP technique is not restricted to any particular problem structure. It can handle nonlinear objective functions and nonlinear constraints. For most reservoir problems, if DP is applied to determine reservoir releases, the state variable is storage, the decision variable is release and the stage is represented by time period.

Hall and Buras (1961) were the first to apply the DP technique in water resources systems analysis. They used DP to solve a problem of capacity allocation among several reservoir sites. Yakowitz (1982) presented state-of-the-art reviews with extensive lists of references on DP and its applications for

several water resources problems and Yeh (1985) did the same for optimal reservoir operation. Models developed for solving reservoir operation problems can be classified by how they characterize the streamflow process. One group of models, called deterministic models, uses a specific sequence of streamflow – either historical or synthetically generated – in deriving operation rules. The other group of models, called stochastic models, uses a statistical description of the streamflow process instead of a specific streamflow sequence.

1.7.1 Deterministic dynamic programming based reservoir operation models

Meier and Beightler (1967) illustrated the applicability of DP in optimizing branched multistage systems in water resources planning. Hall and Shepard (1967) developed a DP-LP technique for optimizing a reservoir system in which the multiple-reservoir system is decomposed into a master-problem and subproblems. The master-problem could be seen as the task to be solved by a system coordinating agency and the subproblems by single-reservoir managers. In that work the subproblems were solved by DP. The schedule of releases and energy production were reported to the system coordinating agency which was modeled by LP.

Larson (1968) introduced the concept of incremental dynamic programming (IDP), putting DP into an iterative context. IDP uses the incremental concept for the state variables. Only a limited state space is considered for a given iteration run. It starts with a feasible initial solution, which can be visualized as a trajectory along the subsequent stages. Traditional DP is then applied in the neighborhood of this trajectory. At the end of each iteration step an improved trajectory is obtained, which is used as the trial trajectory for the next iteration step.

Considering only a limited state space vastly reduces computer time and memory requirements. However, the major setback of using this technique is the possibility of ending up at a local optimum (Turgeon, 1982). That can be avoided by starting with large increments to define the imaginary corridor around the actual trajectory and reducing them gradually as the iteration proceeds. Another way to avoid getting trapped at a local optimum is to repeat the iteration with different initial conditions. Finally, both approaches, i.e., varying increments and different starting solutions, can be coupled (Nandalal, 1986).

Heidari et al. (1971) systematized the use of IDP and referred to it as discrete differential dynamic programming (DDDP). Nopmongcol and Askew (1976) analyzed the difference between IDP and DDDP and concluded that DDDP is a generalization of IDP.

Trott and Yeh (1973) developed a method to determine the optimal planning of a reservoir system with cascade and parallel reservoir configurations. The policy was obtained by decomposing the original problem into a series of subproblems of one state variable each and by applying Bellman's method of successive approximations in such a manner that the series of optimizations over the subproblems converge to a solution of the original problem. Each subproblem was analyzed using the DDDP technique.

Murray and Yakowitz (1979) developed a successive approximation dynamic programming technique using differential dynamic programming principles, constraining a sequential decision variable as applicable to multireservoir control problems in some cases. This approach is known as the constrained differential dynamic programming (CDDP) algorithm.

Karamouz and Houck (1987) formulated two dynamic programming models, one deterministic and one stochastic, to generate operating rules for a single reservoir. The deterministic model comprises a deterministic dynamic program, regression analysis, and simulation. The stochastic model is a stochastic dynamic program. It describes streamflow with a discrete lag-one Markov process. It was concluded that the deterministic model generated rules were effective in the operation of medium to very large reservoirs. The stochastic dynamic programming generated rules were more effective for the operation of small reservoirs.

Harboe (1987) applied DP to a system of reservoirs in which low-flow augmentation was the main purpose. The objective function used in the optimization is to maximize the minimum flow. A sequential optimization starting from upstream and considering one reservoir at a time is employed. The optimum results of one reservoir are used as the inputs to the downstream reservoir. The local optimum obtained was very close to the global optimum due to the high cross-correlation among monthly flows at different locations in the basin.

1.7.2 Stochastic dynamic programming based reservoir operation models

Under real-world conditions the time sequence of the streamflow time series or demands is not known in advance. Therefore, deterministic optimization models are often inadequate for effective water resources systems analysis due to the uncertainties inherent in the prediction of hydrological, economic, and other factors. The stochastic nature of the inflows can be handled by two approaches: implicit or explicit. In the implicit approach, a time series model is used to generate a number of synthetic inflow sequences. The system is optimized for each streamflow sequence and the operating rules are found by multiple regression. During the optimization the synthetic data series are considered as deterministic series.

Although the implicit approach can be easily adopted for single-reservoir optimization, numerous difficulties are encountered in applying it to multireservoir systems. The difficulty of obtaining a computationally manageable algorithm which derives the optimal results becomes much more severe when the streamflows into each reservoir are interdependent. In such a situation, complicated synthetic streamflow-generating models are used to obtain the cross-correlated streamflows into each of the reservoirs. The implicit approach optimizes the system operation under a large number of streamflow sequences, at the expense of computer time. It is therefore employed only for long-range planning purposes.

The explicit approach considers the probability distribution of the inflows rather than specific flow sequences. This approach generates an operation policy comprising storage targets or release decisions for every possible reservoir storage and inflow state in each time step, rather than a mere single schedule of reservoir releases.

Young (1967) proposed an implicit stochastic approach to optimize the operation of a single reservoir. He combined Monte Carlo simulation for synthetic streamflow generation, deterministic DP optimization, and regression analysis to derive the operating strategy which was expressed in terms of release as a function of initial storage volume in the reservoir and inflow during the time step.

Harboe et al. (1970) used deterministic DP to derive the optimal operation policy of a single reservoir serving multiple purposes: water supply, energy generation, flood and water quality control downstream of the reservoir. The last two purposes were considered as maximum storage and minimum downstream release constraints respectively, whereas the target water supply was incorporated as a parameter into the optimization procedure. By varying the level of the water supply target, successive DP optimizations were applied to obtain a family of the optimal operating trajectories with respect to the maximization of the firm energy production. The authors stressed the efficiency of the developed algorithm and suggested that it could easily be implemented as the optimization core of an implicit stochastic DP methodology.

Opricovic and Djordjević (1976) presented an implicit SDP based algorithm for optimal long-term control of a single multipurpose reservoir with both direct and indirect users. The approach takes into account the fact that water already used for one purpose (direct user) can be utilized by another user located further downstream (indirect user). The developed optimization method maximizes the total benefit earned from the delivered water by applying DP at each of the three levels of the adopted hierarchical decomposition of the problem. At the first level, the temporal distribution of reservoir releases is optimized. This is followed by the optimization of the allocation of available releases to direct users in each time interval. At the third level, the release volumes already used by direct users are distributed to indirect users.

Karamouz and Houck (1982) proposed an iterative approach which combined deterministic DP, multiple regression, and simulation to derive a general operating rule for a single water supply reservoir. Although not entirely conforming to the general definition, the method was essentially an implicit stochastic optimization approach. One iterative cycle consisted of deterministic DP optimization over the available historical inflow record, the subsequent derivation of the general linear release rule by means of multiple regression, and the final step which included the simulation of the reservoir operation according to the defined operating rule over a long synthetic sequence of reservoir inflows. The principal idea behind the developed method was to start iterations without any further limitations on the feasible release decision space except those determined by the capacities of the reservoir's outlets and spillways. However, the decision space was narrowed down in DP optimization as the iterations proceeded by using the general release rule defined in the previous cycle. The width of the reduced feasible decision space was corrected by a lower/upper bound factor, the value of which was adjusted at the end of each iteration with respect to the objective function achievement obtained by simulation. The approach has been applied to 48 test cases involving both annual and monthly temporal discretization. In all of the cases, the average annual objective function achievement obtained by simulation over a synthetic inflow record showed improvement over iterations, clearly outperforming the initial iteration outcomes obtained without restricting the release domain.

Karamouz and Houck (1987) derived monthly operating rules for a set of 12 different single-reservoir test cases using their iterative DP model (Karamouz and Houck 1982) and an explicit SDP optimization model. The explicit SDP model used the lag-one Markov chain representation of river flows and the derived optimal operation policy was given in terms of the storage volume at the beginning of the following month as a function of initial storage and inflow at the present time step. The two models were compared based on the objective function achievement derived by simulation over a long synthetic set of river flows. The results of the 12 test cases indicated that the explicit SDP model resulted in better operation policies for smaller reservoirs whereas the iterative DP proved to be more effective for medium to very large reservoirs. Relatively poor performance of SDP on large reservoirs was attributed to

the inability to use finer state discretization as the size of the storage state space was increasing, which would, in turn, impose the well-known dimensionality difficulties associated with SDP.

The sampling stochastic dynamic programming approach (SSDP), first used by Araujo and Terry (1974) for the operation of a hydro system can also be categorized as an implicit stochastic approach. SSDP was used by Dias et al. (1985) for the optimization of flood control and power generation requirements in a multipurpose reservoir. What differentiates SSDP from other implicit stochastic approaches is that the whole set of synthetic 12-month-long streamflow scenarios were simultaneously considered in the optimization process. The approach is said to be very efficient in describing river flow processes and in coupling such a streamflow representation with DP optimization. Kelman et al. (1990) included the best inflow forecast as a hydrological state variable in the SSDP algorithm. In their approach, a historical time series of streamflow forecasts was employed to develop the required conditional probability distributions. Kelman et al. (1988) applied SSDP for planning reservoir operations in a hydroelectric system operated by the Pacific Gas and Electric Company (PG&E) on the North Fork of the Feather River in California, USA.

Faber and Stedinger (2001) compared SSDP models employing the National Weather Service's (NWS) ensemble streamflow prediction (ESP) forecasts to SSDP models based on historical streamflows and snowmelt volume forecasts. The SSDP optimization algorithm, which is driven by individual streamflow scenarios rather than a Markov description of streamflow probabilities, allows the ESP forecast traces to be employed intact, thus taking full advantage of their rich description of streamflow variability and the temporal and spatial interrelationships captured within the traces.

Butcher (1971) used explicit SDP to derive an optimal long-term operating strategy for a single multipurpose reservoir. The optimization model was developed for a monthly temporal discretization assuming that monthly flows were serially correlated. The objective was to maximize the expected annual monetary return gained from irrigation water supply and energy production, and the potential benefit from recreational use of the reservoir. The optimal release policy was expressed as a function of the reservoir state given as the storage volume of the reservoir at the beginning of the month and the inflow during the preceding month.

Loucks et al. (1981) elaborated the explicit SDP approach for the optimization of single-reservoir operation. Stochasticity of inflows represented by the first order Markov chain was explicitly incorporated into the optimization procedure by considering inflows to the reservoir as an additional state variable.

Thus, the procedure assumed a two-state (i.e., reservoir storage at the beginning of and inflow to the reservoir during a time step) variable SDP optimization problem with the decision to be taken being the reservoir storage at the end of a stage. The objective was to minimize the total expected sum of the squared deficit of the release from the respective demand and the squared deviation of the storage from the constant storage target. For each time step, the resulting steady-state operation policy was derived in the form of the final reservoir storage volume as a function of the initial storage and the present inflow. The technique was demonstrated on a simple hypothetical example considering two within-year time periods and a discrete two-class representation for both inflow and reservoir storage variables.

Maidment and Chow (1981) developed two SDP optimization models for a single-reservoir operation problem. The temporal discretization was set to monthly time steps and the authors distinguished between two different representations of inflow stochasticity. One model assumed that the monthly river flows were serially correlated and that the stochasticity of subsequent monthly flow processes was described by inflow transition probabilities (i.e., Markov chain) whereas the second approach considered monthly flows as independently distributed. The objective for both models was to maximize the expectation of the annual net benefit gained from the releases allocated for energy generation and irrigation water supply. The resulting steady-state release strategies were given as a function of the storage volume of the reservoir at the beginning of a month and the inflow during the preceding month.

Stedinger et al. (1984) compared the simulation results based on different operation policies derived for the High Aswan Dam on the River Nile by five SDP based optimization models. Apart from models that used the previous period inflow as the hydrological state variable, the authors proposed approaches that utilized the best forecast of the current period inflow instead. They concluded that the use of the best inflow forecast instead of the inflow during the preceding time period resulted in significant improvements in the operation of the reservoir.

Goulter and Tai (1985) used SDP to model a small hydroelectric system. The variation in the number of stage iterations and the computer time required to reach steady-state conditions with changes in the number of storage states was investigated in this study.

Shrestha (1987) applied SDP to derive optimal operation policies for different configurations of a hydropower system during the planning stage. Simulation of the system operation was carried out based on the SDP based optimum policy to evaluate the system performance. Finally, the optimum

system configuration was selected by comparing the performance values obtained for the different configurations.

Bogardi et al. (1988) investigated the impact of varying the number of storage and inflow classes on the operational performance of SDP for both single and multiple reservoir systems. The results indicated that by simply increasing the number of storage classes beyond certain limits the system performance would not improve much. These results comply with the "law of diminishing returns." Emphasis should rather be placed on the "synchronization" of the number and size of storage and inflow classes, in order to check whether any improvement could be obtained.

Laabs and Harboe (1988) presented three models based on DP, including a deterministic model, an independent probability model, and a Markov model, for finding Pareto-optimal operation rules for a single multipurpose reservoir. In the independent probability model, the inflow probabilities of each time step are considered. Inflow transitional probabilities are considered in the Markov model. The Markov model includes several objective functions and weights for each objective as needed in a compromise programming analysis of multiobjective decision making. A number of Pareto-optimal operation rules were generated. The final selection of the optimal policy can be done only after simulations with these operation rules have been performed and a multiobjective selection criterion applied to the results.

Shrestha et al. (1990) studied the effect of the number of discrete characteristic states and the impact of varying the definition of these characteristic states on SDP model performance. Four real-world cases have been analyzed from different hydrological regimes. It has been found that varying the definition of inflow state discretization renders only marginal changes in the model performance. The factors which could have a direct bearing on the adequate level of storage state discretization are identified as: the hydrological regime in which the system is located (due to the differences in inflow state distribution), the type of system constraints, and the degree of severeness of system constraints.

Huang et al. (1991) compared four explicit SDP optimization models using the operation of the Feitsui Reservoir in Taiwan as a case. The four models were devised upon the assumption that a streamflow process could be modeled as being either serially correlated or independent, and that the consideration of reservoir inflow as an additional state variable could use either the forecast of the present period streamflow or the known observation of the past period flow. Each model was formulated for a 36-period annual cycle and utilized the same objective function, which was to maximize the expectation of the annual energy generation. The authors found that the best performance of the reservoir resulted from the use of the model which assumed serial correlation of river flows, and the previous time step inflow as an additional state variable. However, they recognized that their findings were applicable to the particular case they used, and stressed that the model which used the present time step inflow forecast as a serially correlated hydrological state variable did outperform the other three models when a perfect forecast was assumed available. An additional advantage of SDP models based on the present inflow forecast is that they derive operation policies which specify the optimum achievement of the objective criterion expectation for the given inflow forecast state. Thus, any failure to maintain the optimal operating strategy is due only to the imperfect inflow forecast.

Ratnayake (1995) presented a SDP model to maximize the expected on-peak hydro-energy generation from a reservoir system. The optimization is subject to deterministic constraints on mass balance, maximum and minimum reservoir storage, flood control reserve space in the reservoir, maximum release, and a soft constraint on downstream water requirements for irrigation and salinity intrusion control. The max-min type objective function used is: maximize $\left\{ \underset{R_j}{\text{minimize}}\ \xi\left(E_j\right)\right\}$ where E_j is on-peak energy generation in period j, R_j is release from reservoir during period j and ξ is an operator for calculating expected values over stochastic inflows. The application of the model to reservoirs in the Chao Phraya River system in Thailand resulted in optimistic estimates of the system capability. This is because the calculation is based on the expected value in the objective function, which does not effectively consider the critical drought periods.

A number of studies have dealt with the choice of the hydrological state variable in SDP. For instance, Karamouz and Vasiliadis (1992) used the present time step inflow forecast as an additional state variable in one of their SDP models. In another model, Vasiliadis and Karamouz (1994) adopted both the present period inflow and the next period inflow forecast as hydrological state variables. The latter also applied the Bayes theory to account for the uncertainty of inflow forecasts while updating the inflow transition probabilities during the SDP optimization process. The Bayes–SDP model was found to bring improvement in the operation of the test case as compared to the classical SDP model, which utilized only the present period inflow as a hydrological state variable. In another study, Tejada-Guibert et al. (1995) found that, as compared to deterministic DP or no-hydrological-state-variable SDP models, the operation of the case study system improved if the operation policies were derived by SDP models which used either the present period inflow or the past period inflow in combination with the best forecast of the forthcoming snowmelt runoff as hydrological state

variables. Vedula and Kumar (1996), in their SDP model, utilized both the present period inflow and rainfall forecasts as stochastic state variables.

A quantitative basis for a variety of reservoir operational decisions, from the perspective of both project planning and operation, can be achieved through simulation and optimization methods coupled with stochastic analysis. The most effective strategy for analyzing many reservoir operation problems will involve a combination of both optimization and simulation. Bogardi et al. (1995) developed a model called "ShellDP" based on stochastic dynamic programming and simulation techniques for analyzing multiunit reservoir systems. The model is applicable during both design and operational stages of a reservoir system. The model was successfully used to derive optimal operation policies for the reservoirs in the water supply system of Tunis. Ampitiya (1995) applied the ShellDP package to derive optimal operation policies for reservoirs in the complex Mahaweli water resources scheme in Sri Lanka. In this study, the software was modified to include the objective of hydro-energy generation optimization. Nandalal and Ampitiya (1997), Nandalal (1998), and Nandalal and Sakthivadivel (2002) used this modified model to derive operation policies for reservoirs in several water resource development schemes in Sri Lanka.

1.8 DEVELOPMENTS IN DYNAMIC PROGRAMMING

In stochastic dynamic programming models of reservoir control problems, the continuous state space typically is discretized and the optimal value function is computed at the state space grid points. Values of the optimal value function between these discrete grid points can be obtained by interpolation. Discretization of the state space for high-dimensional problems, which may arise due to the large number of reservoirs in the system or multiple state variables for each reservoir, results in a prohibitively large computational requirement. To overcome this difficulty Fan et al. (2000) developed a regression dynamic programming approach for approximating the optimal value function using penalized regression splines fitted over a subset of the full state space grid. Selection of the subset of state space grid points was made based on statistical methods of experimental design, one of which is Latin hypercube sampling. The regression dynamic programming approach was demonstrated using a hypothetical reservoir system.

For systems with a large number of state variables, e.g., a large number of reservoirs in a system and multiple forecasts on streamflow, the computational requirement of dynamic programming is prohibitive. Fan *et al.* (2001) developed a regression dynamic programming approach to efficiently solve stochastic and nonlinear models of reservoir control problems. The "curse of dimensionality" was alleviated using clever sampling schemes based on statistical methods of experimental design for state space representation. Furthermore, the optimal value function was efficiently approximated using penalized regression splines. Stochastic, multiple-reservoir control problems of two and seven dimensions were solved using the regression dynamic programming approach.

Chandramouli and Raman (2001) developed a dynamic programming based neural network model for optimal multireservoir operation. In the model, multireservoir operating rules were derived using a feed-forward neural network from the results of three state variables' dynamic programming algorithm. The training of the neural network was done using a supervised learning approach with the back-propagation algorithm. A multireservoir system called the Parambikulam Aliyar Project system was used for this study. The performance of the new multireservoir model was compared with the regression-based approach used for deriving the multireservoir operating rules from optimization results and the single-reservoir dynamic programming–neural network model approach. The multireservoir model based on the dynamic programming–neural network algorithm gave improved performance.

Tilmant *et al.* (2002) compared reservoir operation policies obtained from fuzzy and nonfuzzy explicit stochastic dynamic programming. They formulated two models to study the reservoir operation problem for the Mansour Eddahbi Dam in Morocco. The first one is a classical stochastic dynamic programming (SDP) model in which the objective function stresses energy maximization with particular volumes being released for irrigation. The second model is a fuzzy stochastic dynamic programming (FSDP) model in which both hydropower generation and irrigation are considered as fuzzy constraints aggregated by the weighting method. System performance was estimated from simulations based on continuous reoptimization models using the cost-to-go function generated by the SDP algorithm and the membership function generated by the FSDP algorithm. Results indicated that, despite major differences in the mathematical representation of operating objectives and/or constraints, both formulations yielded similar measures of system performance.

Teixeira and Marino (2002) developed a forward dynamic programming model to solve the problem of reservoir operation and irrigation scheduling. The typical scenario for application of the model is composed of a system of two reservoirs in parallel supplying water to as many as three irrigation districts. Two models are coupled. The interseasonal model defines seasonal deliveries from the reservoir system. The intraseasonal model uses area and water allocations generated from the interseasonal model to produce an irrigation schedule for the individual farms in one of the irrigation districts in the reservoir system. Crop evapotranspiration, reservoir evaporation, and inflows are forecast. Upon availability of the current values, the forecast is updated and the model runs to generate a more precise irrigation schedule. This feature permits the application of the model for real-time operation of the irrigation district. At the end of the season, the intraseasonal model is updated. The forward DP model was applied to a real watershed with a planning horizon of two years for the interseasonal model and six months for the intraseasonal model.

Reservoir operation involves a complex set of human decisions depending upon hydrological conditions in the supply network, including watersheds, lakes, transfer tunnels, and rivers. Water releases from reservoirs are adjusted in an attempt to provide a balanced response to different demands. When a system involves more than one reservoir, computational burdens have been a major obstacle in incorporating uncertainties and variations in supply and demand. A new generation of stochastic dynamic programming was developed in the 1980s and 1990s to incorporate the forecast and demand uncertainties. The bayesian stochastic dynamic programming (BSDP) model and its extension, the demand driven stochastic dynamic programming (DDSP) model, are among those models (Karamouz and Mousavi, 2003). Recently, a fuzzy stochastic dynamic programming model (FSDP) (Mousavi *et al.*, 2004) was also developed for a single reservoir to model the errors associated with discretizing the variables using fuzzy set theory. In this study the DDSP and the FSDP models were extended and simplified for the complex system of the Dez and Karoon Reservoirs in the southwestern part of Iran. The simplified models are called condensed demand driven stochastic programming (CDDSP) and condensed fuzzy stochastic dynamic programming (CFSDP). The optimal operation policies developed by the CDDSP and the CFSDP models were simulated in a classical model and a fuzzy simulation model, respectively. The case study was used to demonstrate the advantages of implementing the proposed algorithm, and the results show the significant value of the proposed fuzzy based algorithm.

Kumar and Baliarsingh (2003) presented a new algorithm, folded dynamic programming, to overcome the curse of dimensionality inherent in dynamic programming. Incremental DP, discrete differential DP, DP with successive approximation, and incremental DP with successive approximation are some of the algorithms evolved to tackle this curse of dimensionality for DP. But in all these cases it is difficult to choose an initial trial trajectory, to get at an optimal solution, and there is no control over the number of iterations required for convergence.

The proposed folded DP, an iterative process, does not need an initial trajectory to start and thus overcomes these difficulties. In the folded DP algorithm, the entire storage state space at each time period is divided into four equal state increments to form five grid points. The developed algorithm was applied to a hypothetical reservoir system. Operation policy obtained using the folded DP algorithm compared well with that of the IDP algorithm.

Umamahesh and Chandramouli (2004) presented a fuzzy dynamic programming (FDP) model developed for optimal operation of a multipurpose reservoir. The model is applied to derive the ten-daily operation policy of the Hirakud Reservoir on the River Mahanadi in India. The objectives of the reservoir, namely, irrigation, hydropower, and flood control are considered as fuzzy. The objective function of the FDP model is to maximize the minimum expected satisfaction level of the fuzzy objectives. The level of satisfaction of an objective (membership grade) is a function of reservoir release for irrigation and hydropower, and initial reservoir storage for flood control. The reservoir is simulated using the operation policy derived from the FDP model and the performance of the reservoir is evaluated. The model not only considers uncertainty due to the variability of inflows, but also considers the uncertainty caused by imprecisely defined objectives.

Mousavi et al. (2004) introduced a new way of incorporating fuzzy logic concepts to better capture and manage some uncertainties in applying SDP formulations for reservoir operation. Their model, which is called fuzzy-state stochastic dynamic programming (FSDP), takes into account both uncertainties due to the random nature of hydrological variables and imprecision due to variable discretization. In the model, fuzzy transition probabilities for stochastic hydrological state variables are calculated by defining a fuzzy Markov chain. These fuzzy probabilities are derived based on the fuzzy frequency concept considering different frequencies for different points of a class interval. To show the effectiveness of the proposed method, FSDP was applied to the Zayandeh-Rud river–reservoir system in Isfahan, central Iran. A comparison with a demand-driven stochastic dynamic programming model shows the robustness of the FSDP solutions with respect to the type of discretization scheme used in calculating the transition probabilities.

2 Incremental dynamic programming in optimal reservoir operation

Very often reservoirs are built and operated to satisfy water quantity requirements such as irrigation or drinking water consumption, hydropower production, etc. Operation of these reservoirs is vital to maximize benefits and DP in various forms can be applied for this purpose. This chapter illustrates the applicability of several different forms of incremental dynamic programming (IDP) in the derivation of optimal operation patterns for different reservoir systems.

2.1 IDP IN OPTIMAL RESERVOIR OPERATION: SINGLE RESERVOIR

Dynamic programming can be applied to derive optimal operation policies for a single reservoir. Use of conventional DP in this task is one possibility. This section shows the application of IDP for two reservoir systems: (a) the Kariba Reservoir in Zambia and Zimbabwe, and (b) the Ubol Ratana Reservoir in Thailand, based on the analysis of Budhakooncharoen (1986, 1990). The IDP technique, which considers only a limited state space at a time, has lower computer storage requirements.

2.1.1 Application of IDP to the Kariba Reservoir

The Kariba Reservoir is a single-unit water resource system that utilizes the water resource of the Zambezi River to produce hydroelectric energy. The Zambezi River rises in northern Zambia. After flowing through Angola, it forms the boundary between Zambia and Zimbabwe. Finally, it passes through Mozambique to discharge into the Indian Ocean north of Beira. Figure 2.1 shows the Zambezi River basin. The catchment area of the basin up to the Kariba Dam is about $665\,600\,\mathrm{km}^2$. The Kariba Reservoir is shared and operated by CAPCO (Central African Power Corporation).

The Kariba project was completed in two stages. In the first stage, a dam across the Kariba Gorge on the Zambezi River

was constructed with a 608 MW underground hydroelectric power station on the south bank. The power station started operations in March 1962. In the second stage, a 592 MW underground power plant was constructed on the north bank bringing the total installed capacity to 1200 MW. The second stage was completed in 1977. Power from the Kariba hydropower stations is supplied to Zambia and Zimbabwe, which share equally the available generating power and cost of operation. Salient features of the dam and reservoir are given in Table 2.1. Characteristic curves for the Kariba Reservoir are given in Figure 2.2.

The adopted operational rule curve of the Kariba Reservoir in Figure 2.3 shows the sequence of preferable water levels to be kept in different time periods in the annual cycle (Budhakooncharoen, 1990). According to it, the reservoir should be pre-emptied in February in order to accommodate the expected flood flows during March and April.

The study was carried out based on Zambezi River flow data at the Kariba Reservoir from October 1961 to September 1985. Mean annual flow at this location is $54\,689 \times 10^6\,\mathrm{m}^3$. During this period, a minimum annual inflow of $23\,729 \times 10^6\,\mathrm{m}^3$ in the hydrological year 1972 and a maximum annual inflow of $97\,902 \times 10^6\,\mathrm{m}^3$ in the hydrological year 1977 were observed.

The optimum operation of the Kariba Reservoir to maximize hydro-energy generation was developed based on the IDP technique.

IDP MODEL FORMULATION
Formulation of the IDP model for a system comprising a single reservoir with hydropower generation facility shown in Figure 2.4 is given below.

The system is operated on a monthly basis and the reservoir begins and ends its operation cycle with a given amount of water stored. The forward algorithm of dynamic programming is used in the optimization.

Table 2.1. *Salient features of the Kariba dam, reservoir, and power house*

Description	
Reservoir	
Normal high-water level	488.5 m above Kariba datum
Normal storage capacity	$64\,880 \times 10^6\,\mathrm{m}^3$
Minimum water surface level	475.5 m above Kariba datum
Minimum storage capacity	$50 \times 10^6\,\mathrm{m}^3$
Dam	
Type	Double curvature/mass concrete
Height	128 m
Length at crest	617 m
Width at crest	24.4 m
Spillway	
Type	Radial gate
Gate	6 Nos. 9.1 m × 9.45 m
Maximum discharge capacity	$9400\,\mathrm{m}^3/\mathrm{s}$
Power house	
Average net head	86 m
Turbine discharge	$277.6\,\mathrm{m}^3/\mathrm{s}$
Turbine	12 Nos. Francis
Installed capacity	1200 MW

OBJECTIVE FUNCTION

The objective function of the IDP model is to maximize total hydro-energy generation over a prespecified time period:

$$OF = \text{Maximize} \sum_{j=1}^{T} E_j, \qquad (2.1)$$

where

$E_j = 9.81 \eta Q_j H_j t_j / 10^6$ (MWh),

EL_j = elevation of reservoir water level during period j (m),

$H_j = EL_j - TWL$ (m),

h_j = reservoir water level at beginning of period j (m above datum),

Q_j = discharge through turbine during period j (m^3/s),

TWL = tail water level (m above datum),

t_j = time in period j (h), and

η = overall generation efficiency (= 0.75).

STAGES, STATE, AND DECISION VARIABLES

The state of the system is described by water available in the reservoir at the beginning of any time step. Consecutive time steps are identified as stages. The decision variable is water released from the reservoir. The maximization is subjected to constraints in storage volume and release.

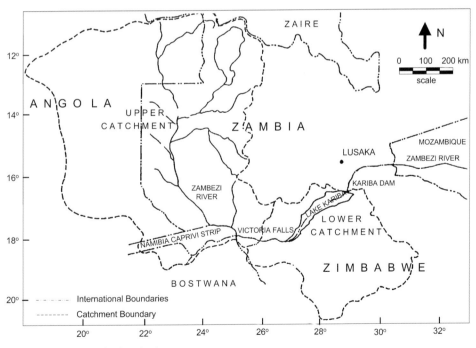

Figure 2.1 Kariba Reservoir and Zambezi River basin

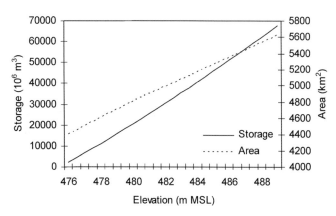

Figure 2.2 Characteristic curves of the Kariba Reservoir

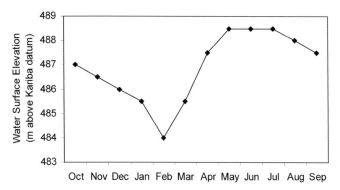

Figure 2.3 Rule curve of the Kariba Reservoir

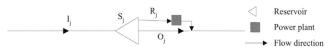

Figure 2.4 Single-reservoir configuration

STORAGE VOLUME CONSTRAINT

The system is operated on a monthly basis. Since operation policy is derived for annual cycles,

$$S_1 = S_{T+1}, \tag{2.2}$$

where

S_1 = storage volume at beginning of first period (first month) $(10^6\,\text{m}^3)$,

S_{T+1} = storage volume at end of last period (last month) $(10^6\,\text{m}^3)$, and

T = total number of time steps (months).

For all other months reservoir storage belongs to the set of admissible storage volume:

$$S_{\min} \le S_j \le S_{\max}, \tag{2.3}$$

where

S_j = storage volume at beginning of period j $(10^6\,\text{m}^3)$,

$S_{\min}$ = allowable minimum storage volume $(10^6\,\text{m}^3)$, and

$S_{\max}$ = allowable maximum storage volume $(10^6\,\text{m}^3)$.

RELEASE CONSTRAINT

The capacity of hydropower generators sets a maximum limit to reservoir release. Since a minimum release request is not considered, the minimum release is set to zero. The release during any month should be within this feasible release range:

$$0 \le R_j \le R_{j,\max}, \tag{2.4}$$

where

R_j = reservoir release during period j $(\approx Q_j)$ $(10^6\,\text{m}^3)$, and

$R_{j,\max}$ = maximum allowable release through turbines in period j $(10^6\,\text{m}^3)$.

STATE TRANSFORMATION EQUATION

The state transformation equation based on the principle of continuity is as follows:

$$S_{j+1} = S_j + I_j - E_j - R_j - O_j, \tag{2.5}$$

where

E_j = evaporation from reservoir during period j $(10^6\,\text{m}^3)$,

I_j = inflow to the reservoir during period j $(10^6\,\text{m}^3)$, and

O_j = spillage water during period j $(10^6\,\text{m}^3)$.

O_j = Max $[S_j + I_j - E_j - R_j - S_{\max}, 0]$.

Other variables are as defined before.

RECURSIVE EQUATION

The deterministic backward optimization procedure begins at some known point in the future and proceeds backward in time to the present.

The recursive equation of deterministic DP for estimating the maximum energy output is

$$F_j^*(S_j) = \underset{R_j}{\text{Max}}\Big[E_j(S_j, R_j) + F_{j+1}^*(S_{j+1})\Big], \tag{2.6}$$

where maximum energy output at stage j associated with the particular state S_j, given the suboptimal energy generation from future stages, $F_{j+1}^*(S_{j+1})$. The decision R_j identified in the course of maximization transforms the system from state S_j into state S_{j+1}.

The model was run for four different cases, an average year (1966), two dry years (1972 and 1981), and a wet year (1968). The operation over a 23-year period (1962–84) is also included.

EFFECT OF INITIAL CORRIDOR WIDTH ON CONVERGENCE BEHAVIOR

The model was run for six different initial corridor widths. In all of these cases, the procedure of IDP started with the same

Table 2.2. *Effect of initial corridor width: Kariba Reservoir*

Initial half corridor width ($10^6\,\mathrm{m}^3$)	Optimal energy generation for water years 1962–84 ($10^6\,\mathrm{MWh}$)	Difference from maximum value (%)	Number of iterations for converging to optimal solution
100	208.72	0.80	324
200	208.71	0.81	173
300	208.69	0.82	127
1000	210.41	0.00	55
2000	207.40	1.43	33
3000	200.45	4.73	16

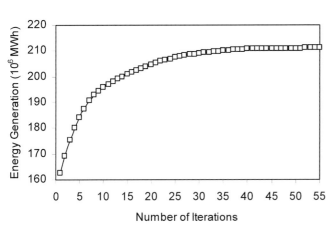

Figure 2.5 Rate of convergence in IDP for initial half width of $1000 \times 10^6\,\mathrm{m}^3$

Table 2.3. *Maximum energy generation: Kariba Reservoir*

Water year	Installed capacity of hydropower plant (MW)	Maximum energy generation (MWh)
1966 (average year)	600	4 730 400
1968 (wet year)	600	4 730 400
1972 (dry year)	600	4 730 400
1981 (dry year)	1200	6 977 739
1962–84	1200	210 408 600

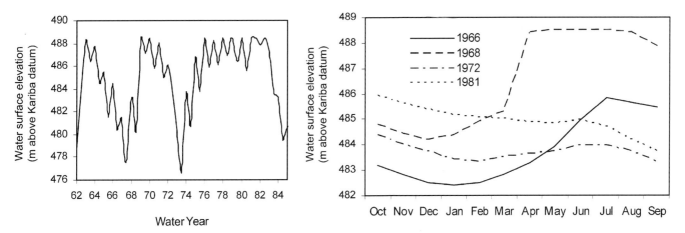

Figure 2.6 Optimal operations to maximize energy generation of the Kariba Reservoir by IDP

initial trial trajectory. The historical record of the Kariba Reservoir operation for water years from 1962 to 1984 was heuristically selected as the initial trial policy.

The total 23-year energy generation and the number of iterations of IDP to converge to the optimal results for different initial corridor widths are given in Table 2.2.

These results indicate that the initial corridor width has not much effect on the optimal result. It has a rather high effect on the rate of convergence to the optimum point. The larger the initial corridor width, the fewer the iterations required to converge to the optimal solution. The rate of convergence to the optimal result for an initial half corridor width of $1000 \times 10^6\,\mathrm{m}^3$ is shown in Figure 2.5.

MAXIMUM ENERGY GENERATION

Maximum energy generation was obtained based on the single objective function of maximizing energy output without considering the actual energy demand. This amount of energy production indicates the maximum hydropower potential of the system.

It is revealed that the optimal energy outputs obtained for different initial corridor widths do not vary significantly. In order to reduce the number of iterations to converge to the optimal solution, the initial corridor width of $1000 \times 10^6 \, \text{m}^3$ was heuristically applied. This version also gave the best solution compared to the others as shown in Table 2.2.

Figure 2.6 displays the optimal operation of the reservoir to maximize hydropower output by applying the historical operation as the initial trial trajectory. The optimal energy outputs are summarized in Table 2.3.

Table 2.4. *Salient features of the Ubol Ratana dam, reservoir, and power house*

Description	
Reservoir	
Normal high-water level	182.0 m MSL
Normal storage capacity	$2264 \times 10^6 \, \text{m}^3$
Minimum water surface level	175.0 m MSL
Minimum storage capacity	$502 \times 10^6 \, \text{m}^3$
Maximum flood level	186.6 m MSL
Water surface area	401 km^2
Dam	
Type	Rockfill/clay core
Height	35.1 m
Elevation at crest	188.1 m
Length at crest	855 m
Width at crest	6 m
Maximum width (at base)	125 m
Spillway	
Type	Radial gate (Orifice)
Gate	4 Nos. 12.0 m × 7.8 m
Maximum discharge capacity	3500 m^3/s
Sill level	171 m MSL
Penstock	
Number	3 units
Dimensions	4.5 m × 4.5 m
Maximum discharge	210 m^3/s
Power house	
Type of power house	Underground
Type of turbines	Kaplan, vertical shaft
Number of turbines	3 units
Installed capacity	8.3 MW
Design head	16 m

2.1.2 Application of IDP to Ubol Ratana Reservoir

The multipurpose Ubol Ratana Project uses the water resources of the Nam Pong River in Thailand to generate hydroelectric energy and supply irrigation water. It protects downstream areas from floods while used for fishery, navigation, and recreation.

The Ubol Ratana Dam is built across the Nam Pong River at Tambon Kok Soong, Amphoe Ubol Ratana, approximately 50 km north of the city of Khon Kaen as shown in Figure 2.7. Its construction was completed in 1965. The total catchment area above the dam is about 14 000 km^2.

The salient features of the dam, reservoir, and power house are summarized in Table 2.4. Characteristic curves of the Ubol Ratana Reservoir are shown in Figure 2.8.

Under the Ubol Ratana Project, the Nong Wai irrigation system distributes the water released through the hydropower plant for cultivation on both banks of the Nam Pong River. The Nong Wai diversion weir shown in Figure 2.7 is of ogee shape. It is a reinforced concrete weir of height 5.9 m and crest length 125.4 m. The weir height can be augmented by 0.6 m by pumping air into a rubber weir fixed onto the top of the concrete weir. The cultivation area on the left bank is about 19 580 ha while that on the right bank is about 22 000 ha. The flood mitigation system of the Nong Wai irrigation project consists of an emergency spillway, 3.5 m high and 240 m long. This spillway is able to release a discharge of 1000 m^3/s. The reinforced concrete weir (Nong Wai weir) itself is able to release a maximum of 1500 m^3/s.

The rule curve of the Ubol Ratana Reservoir derived by a simulation technique is shown in Figure 2.9. The rule curve shows the minimum water level that should be maintained at the end of each calendar month.

Water above the rule curve level is released for provision of maximum flood retention volume. If the reservoir level is below or at the rule curve level, only the water to satisfy downstream requirements is recommended to be released as long as the resource is available.

If the reservoir level rises above the rule curve level, all inflowing water is released at a rate up to the permissible downstream flow of 400 m^3/s unless the operation rules for a flood event have to be applied. As far as the conditions such as head, capacity of the penstock and the turbines allow, all water release is utilized to generate electric energy.

The study was based on an observed flow record from 1966 to 1988. Mean annual inflow in the river is $2240 \times 10^6 \, \text{m}^3$ with a minimum observed annual inflow of $850 \times 10^6 \, \text{m}^3$ in the hydrological year 1981 and a maximum annual inflow of $5900 \times 10^6 \, \text{m}^3$ in hydrological year 1978. Downstream flooding could not be avoided in this year.

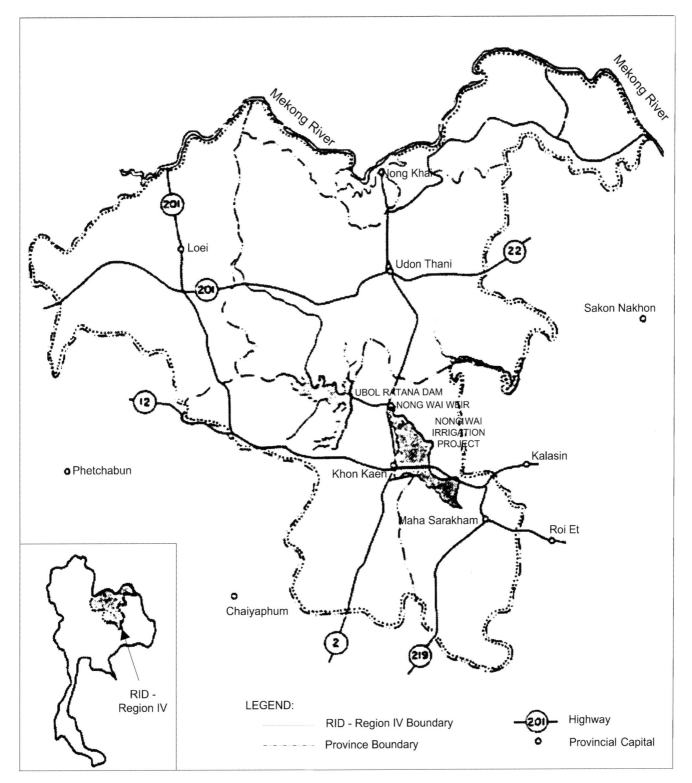

Figure 2.7 Ubol Ratana Reservoir system

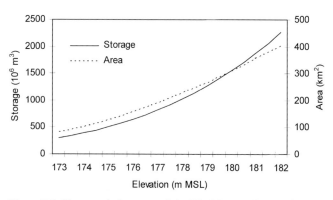

Figure 2.8 Characteristic curves of the Ubol Ratana Reservoir

Table 2.5. *Effect of initial corridor width: Ubol Ratana Reservoir*

Initial half corridor width ($10^6 \, m^3$)	Optimal energy generation for water years 1966–1988 (MWh)	Number of iterations for converging to optimal solution
10	1 558 831	147
20	1 559 663	80
30	1 559 304	55
40	1 558 684	52
80	1 559 361	49

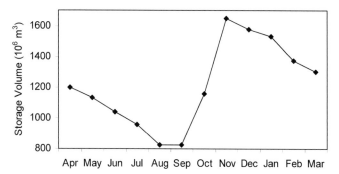

Figure 2.9 Rule curve of the Ubol Ratana Reservoir

The model presented for the Kariba Reservoir is also used for the derivation of the optimal operation pattern for the Ubol Ratana Reservoir. The model was run for four different hydrological years, average year (1971), wet year (1978), dry year (1981) and long-term operation from water year 1966 to 1988.

EFFECT OF THE INITIAL CORRIDOR WIDTH ON THE CONVERGENCE BEHAVIOR

The IDP model is applied to derive the optimal operation to maximize the total energy generation of the Ubol Ratana hydropower plant. The historical record of the Ubol Ratana Reservoir operation served heuristically as the initial trial trajectory in the IDP process. The improved trajectory is sought within the range of a predefined corridor for each iteration.

To study the effect of initial corridor width upon the convergence behavior of IDP, the model was run for five different initial corridor widths starting with the same initial trial trajectory. The historical record of the Ubol Ratana Reservoir operation from water year 1966 to 1988 was considered as the initial trial policy.

Total energy production over the period of 23 years and the number of iterations of IDP to converge to the optimal result for different initial corridor widths are shown in Table 2.5.

The effect of initial half corridor width on the convergence behavior of IDP in the case of the Ubol Ratana Reservoir is similar to that of the Kariba Reservoir. From these results, it is obvious that the initial corridor width does not affect the optimal value of the objective function. However, it affects the rate of convergence to the optimum point. The larger the initial corridor width, the smaller the number of iterations required to converge to the optimal solution.

MAXIMUM ENERGY GENERATION

The optimal energy outputs of the Ubol Ratana hydropower plant obtained for different initial corridor widths of IDP are almost identical. To reduce the computation effort to achieve the optimal solution, the initial corridor width of $80 \times 10^6 \, m^3$ was applied subsequently. This corresponds to approximately 3.13% of the normal storage capacity, $2559 \times 10^6 \, m^3$.

The optimal operation policies to maximize hydropower output of the Ubol Ratana Reservoir by applying historical operation as the initial trial trajectory are illustrated in Figure 2.10. The optimal solutions were derived based on the single objective of maximizing energy generation without considering the other objectives of the reservoir system, with the exception of the flood control constraint by preventing the water level rising beyond 182.0 m MASL. The optimal energy outputs are summarized in Table 2.6.

The energy production of the Ubol Ratana Reservoir obtained with the IDP model is significantly higher than that of the historical record. This is because historical operation was carried out without an optimization technique, which relies on the knowledge of inflows and demands imposed on the operation.

2.1.3 Applicability of IDP for a single reservoir

The two examples show the applicability of the IDP technique in the derivation of an optimum operational strategy for a system comprising a single reservoir. The possibility to reduce

Table 2.6. *Maximum energy generation: Ubol Ratana Reservoir*

Water year	Maximum energy generation (MWh)	Percentage of wet year	Historical energy generation (MWh)	Percentage of wet year
1971 (average year)	88 305	73	82 828	69
1978 (wet year)	120 274	100	93 988	78
1981 (dry year)	42 676	35	34 517	29
1966–88	1 559 361	—	1 143 552	—
Mean annual (1966–88)	67 798	56	49 720	41

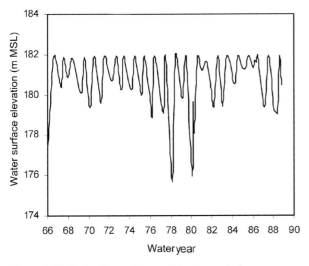
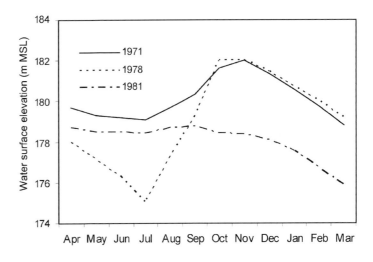

Figure 2.10 Optimal operation policies to maximize energy generation of the Ubol Ratana Reservoir

the corridor width in successive iterations finally results in a small discretization step size for the state variable. This is an advantage of the IDP technique since solution of the same optimization problem with that discretization step size using the conventional DP technique would require a considerable amount of computer storage and time.

2.2 IDP IN OPTIMAL RESERVOIR OPERATION: MULTIPLE-RESERVOIR SYSTEM

In a multiple-reservoir system, it is necessary to obtain an operation policy for all the reservoirs simultaneously, because the optimum condition of the system cannot be investigated by considering reservoirs in isolation. This requirement, which implies the "curse of dimensionality," prohibits the use of conventional DP for a multiple-reservoir system. The IDP technique is considered to be a suitable method to overcome the high dimensionality problem.

APPLICATION OF IDP TO THE MAHAWELI WATER RESOURCES SYSTEM

Application of the IDP technique to derive optimal operation policies for a multiple-reservoir system is presented in this section based on the work of Nandalal (1986).

MAHAWELI WATER RESOURCES SYSTEM

The Mahaweli Water Resources Development Scheme is a multipurpose water resources scheme that harnesses the hydroelectric and irrigation potential of the Mahaweli Ganga (River) in Sri Lanka. The scheme comprises a complex network of regulating reservoirs and diversion structures built on the main stem of the Mahaweli River as well as on its tributaries and diversion routes.

As the schematic diagram of the Mahaweli system in Figure 2.11 illustrates, there are three reservoirs on the main stem of the Mahaweli River namely the Victoria, Randenigala and Rantembe Reservoirs. Each of these reservoirs has a power plant. These reservoirs serve the purposes of power generation and flow regulation for irrigation. The Caledonia

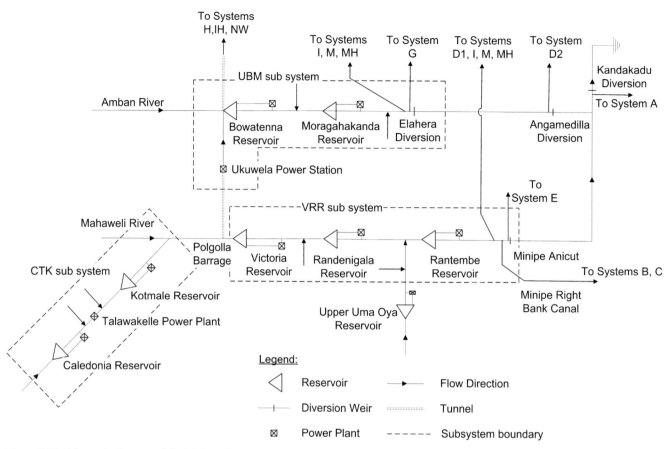

Figure 2.11 Schematic diagram of the Mahaweli system

Reservoir, Talawakelle power plant, and Kotmale Reservoir are located on Kotmale Oya (Creek), a major tributary of the Mahaweli River. Downstream of the Kotmale Reservoir is the Polgolla Barrage, which plays a vital role in this water resources system. It is used for an interbasin water transfer from the Mahaweli River to the adjacent Amban Ganga Basin via a diversion tunnel. The diverted water is used to generate power at a power station at Ukuwela before being collected in the Bowatenna Reservoir. The Bowatenna Reservoir is used as a regulating reservoir for diverting irrigation water to irrigation systems H, IH, and NW while serving the purpose of power generation by downstream discharges. The Moragahakanda Reservoir, which is located downstream of the Bowatenna Reservoir, is also a multipurpose structure that provides hydropower generation and flow regulation for irrigation. Major diversion points of the system are Polgolla, Bowatenna, Elahera, Angamedilla, Minipe, and Kandakadu. The Minipe Anicut (weir) diverts water to both the right and left bank canals in order to fulfil the requirements of systems B, C, and E. A canal from Minipe is envisaged to feed system D1 also from the water available at Minipe. The features of the

components of the Mahaweli water resources system are given in Table 2.7.

IDP MODEL FORMULATION

The IDP model formulated for a serially linked two-reservoir system, a subsystem of the Mahaweli system is shown in Figure 2.12. Due to the small size of the Rantembe Reservoir, the Rantembe power plant is assumed to be a run-of-the-river power plant (Nandalal, 1986).

The system is operated on a monthly basis and the reservoirs begin and end their yearly operational cycle with a given amount of water stored. The forward algorithm of dynamic programming is used in the optimization.

OBJECTIVE FUNCTION

The objective function is to maximize the total energy generation from the three power plants over a pre-specified time period. That is:

$$OF = \text{Maximize} \sum_{j=1}^{N} \text{TEP}_j, \qquad (2.7)$$

Table 2.7. *Principal features of the existing and proposed reservoirs/power plants*

Item		Caledonia	Talawakelle	Kotmale	Victoria	Randenigala	Rantembe	Ukuwela	Bowatenna	Moragahakanda
Hydrology										
Catchment area	km^2	235.0	363.0	562.0	1891.0	2365.0	3111.0	1292.0	506.0	782.0
Average annual discharge	10^6 m^3	412.0	636.0	985.0	1984.0	2528.0	3126.0	2133.0	1343.0	968.0
Reservoir										
Extreme max. WL	m	1363.5	1200.0	704.3	441.2	236.2	155.0	446.4	252.8	195.6
Normal max. WL	m	1360.0	1200.0	703.0	438.0	232.0	152.0	440.8	251.8	195.0
Min. operating WL	m	1341.0	1193.0	665.0	370.0	203.0	140.0	438.4	243.8	170.0
Storage capacity										
Normal max. WL	10^6 m^3	45.7	2.6	172.9	720.0	857.0	22.0	4.1	52.0	902.8
Min. operating WL	10^6 m^3	15.7	0.6	22.2	34.0	295.0	4.4	2.0	17.1	217.2
Design spillway discharge	m^3/s	2470.0	3500.0	5560.0	7900.0	8085.0	10235.0	—	4340.0	3400.0
Low level outlet capacity	m^3/s			133.0	760.0	200.0	180.0	1.0	1.0	(100.0)
Dam										
Type of dam		Concrete	Concrete	Rockfill	Concrete	Rockfill	Concrete	Concrete	Concrete	Rockfill
Crest length	m	270.0	102.0	600.0	520.0	485.0	415.0	144.0	226.0	—
Height	m	70.0	20.0	87.0	122.0	94.0	43.5	14.6	30.0	—
Hydraulic turbine										
Number of units		1	3	3	3	2	2	2	1	2
Type of turbine		Francis	Francis	Francis	Francis	Francis	Francis	Francis	Francis	Francis
Rated power	MW	44	3 × 68	3 × 67	3 × 70	2 × 63	2 × 24.5	2 × 19	40.0	2 × 12.5
Rated head	m	144.0	468.0	201.5	190.0	78.0	31.5	78.0	52.7	54.8
Discharge	m^3/s	35.0	50.0	3 × 38.0	3 × 46.7	2 × 90.0	2 × 90.0	2 × 28.3	94.9	56.6

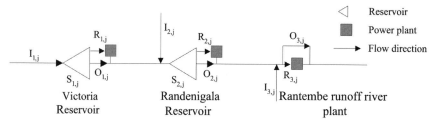

Figure 2.12 System configuration: Victoria, Randenigala, Rantembe subsystem

where

$$\mathrm{TEP}_j = \sum_{i=1}^{3} 9.81 \times \eta \times Q_{i,j} \times H_{i,j} \times t_j/10^6 \ (\mathrm{MWh}), \text{ total energy production,}$$

$\mathrm{EL}_{i,j}$ = elevation of reservoir i during period j (m),

$H_{ij} = \mathrm{EL}_{i,j} - \mathrm{TWL}_{i,j}$ (m),

N = number of periods (= 12, months in a year),

$Q_{i,j}$ = release from reservoir i during period j (m³/s),

$\mathrm{TWL}_{i,j}$ = tailwater level of power station i during period j (m),

t_j = time in period j (h), and

η = overall efficiency = 0.75 (turbines + generators + transmission).

Reservoir storages and releases are assumed to be the state variables and decision variables, respectively. This maximization is subject to constraints on reservoir storages and releases.

Storage constraints and release constraints are identical with those for the single-reservoir case (Eq. 2.3 and Eq. 2.4). State transformation equations, which are expressed by the principle of continuity, are as follows.

For the upstream reservoir,

$$S_{1,j+1} = S_{1,j} + I_{1,j} - E_{1,j} - R_{1,j} - O_{1,j}. \tag{2.8}$$

For the downstream reservoir,

$$S_{2,j+1} = S_{2,j} + I_{2,j} - E_{2,j} - R_{2,j} + R_{1,j} + O_{1,j} - O_{2,j}. \tag{2.9}$$

For the run-of-the-river plant,

$$R_{3,j} = R_{2,j} + I_{3,j} + O_{2,j} - O_{3,j}. \tag{2.10}$$

Where

$E_{i,j}$ = evaporation from reservoir i during period j, $i = 1, 2$ (10⁶ m³),

$I_{i,j}$ = incremental inflow to reservoir i during period j, $i = 1, 2$ (10⁶ m³),

$I_{3,j}$ = incremental inflow to runoff river plant (10⁶ m³),

$O_{i,j}$ = spill from reservoir i during period j, $i = 1, 2$ (10⁶ m³),

$O_{3,j}$ = spill over run-of-the-river plant (10⁶ m³),

$R_{i,j}$ = release from reservoir i during period j, $i = 1, 2$ (10⁶ m³),

$R_{3,j}$ = release through run-of-the-river plant (10⁶ m³), and

$S_{i,j}$ = storage of reservoir i at beginning of period j, $i = 1, 2$ (10⁶ m³).

In the DP formulation of this problem there exist 12 stages, a state vector S_j having two values S_{ij}, and a decision vector R_j having two values $R_{i,j}$, whereas $R_{3,j}$ is defined as a function of inflows and upstream releases.

The DP recursive equation, which is used to determine the deterministic optimum solution within each corridor, can be expressed as

$$F_{j+1}^*(S_{j+1}) = \underset{R_j}{\mathrm{Max}} \left[\mathrm{TEP}_j(S_j, S_{j+1}) + F_j^*(S_j) \right]; \qquad j = 1, 2, \ldots, N, \tag{2.11}$$

where $F_{j+1}^*(S_{j+1})$ is the maximum total of the objective function value from stage 1 to stage $j + 1$, when the state at stage $j + 1$ is S_{j+1}.

CONSTRUCTION OF CORRIDORS FOR A TWO-STATE VARIABLE IDP MODEL

A corridor composed of three values of the state variable is constructed around the initial trajectory whenever possible. In general, the corridor is defined symmetrically around the trial trajectory of state variables as described in the following. For a two-reservoir system, the state of the system in stage j is defined by the storage volumes of the two reservoirs at the beginning of the period j (S_{1j}, S_{2j}). Then the three boundary points of the corridor with regard to S_{1j} can be defined as ($S_{1j} - \Delta_1$), S_{1j}, and ($S_{1j} + \Delta_1$). Similarly, the three boundary points for S_{2j} can also be defined as ($S_{2j} - \Delta_2$), S_{2j}, ($S_{2j} + \Delta_2$), where Δ_1 and Δ_2 are the half corridor widths for state variables 1 and 2 respectively. These imply the identification of nine points in the two-dimensional storage space as shown in Figure 2.13. However, asymmetrical corridors may result if the boundaries of the corridors exceed the minimum or maximum limits of live storage capacities. Larger corridor widths are used for the initial cycles, which ensures that the optimal trajectories are obtained within a small number of iterations. Since the initial trajectory for any later cycle is the optimal trajectory for its preceding cycle and thus closer to optimality than the initial one, smaller corridor widths can be used for later cycles to search for the

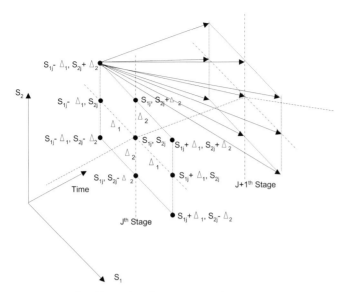

Figure 2.13 Corridor points for two-reservoir case

Table 2.8. *Effect of initial corridor width in IDP*

Half corridor width ($10^6 \, \text{m}^3$)	Annual energy generation (MWh)
30	1 464 990
40	1 465 174
50	1 465 064

optimal trajectory. In this study, the corridor widths were halved after each cycle.

After the construction of a corridor around the trial trajectory, the optimal trajectory and the corresponding objective function value within the corridor are sought. This is done by means of a conventional dynamic programming algorithm, but restricting the computations of the state transformations to only those values of the state variables defined by the corridor. The calculation steps in the procedure are presented in Figure 2.14.

TESTS FOR CONVERGENCE
As indicated previously, the optimal trajectory for a given corridor width is obtained iteratively. The improvement of the return from trajectories of subsequent iterations decreases as the iterations progress. The convergence criterion can be expressed as

$$\delta_i = \frac{\left| F_i^* - F_{i-1}^* \right|}{\left| F_1^* - F_0^* \right|}; \qquad i = 1, 2, \ldots, I, \qquad (2.12)$$

where
F_i^* = return from optimal trajectory for the ith iteration of given cycle ($i = 0, 1, 2, \ldots$), and
I = maximum number of iterations per cycle.

If, during any of the intermediate cycles, the iterative process yields a value of δ_i which does not represent a significant improvement in the return, that is

$$\delta_i \leq \varepsilon; \qquad i = 1, 2, \ldots, I, \qquad (2.13)$$

the computational cycle will be terminated. The next cycle starts with a smaller (half-size) corridor considered around the optimal trajectory of the completed cycle. After the final iteration of each cycle, the following test will be made in order to determine the convergence of the algorithm toward the optimal solution.

$$\lambda \geq \frac{\left| F_j^* - F_{j-1}^* \right|}{F_{j-1}^*}, \qquad (2.14)$$

where
F_j^* = return from the optimal trajectory for the jth cycle ($j = 1, 2, 3, \ldots$).
λ is an arbitrary convergence criterion which terminates the IDP procedure once the above criterion is satisfied. The trajectory which yields the optimum return is identified as the solution of the optimization problem. In the present study, ε and λ were assigned the values of 0.001 and 0.0001 respectively.

APPLICATION OF THE IDP MODEL TO THE MAHAWELI WATER RESOURCES SYSTEM
The model was run for the average year (average of inflows for 32 years) for the Victoria, Randenigala, and Rantembe subsystem.

EFFECT OF THE INITIAL CORRIDOR WIDTH
The model was run for three different initial corridor widths. The same initial trial trajectory was assumed in all these cases. The initial trial trajectories and the resultant optimal trajectories for the Victoria Reservoir and the Randenigala Reservoir are presented in Figure 2.15 and the total annual energy generation is given in Table 2.8. The rates of convergence to the optimal result are shown in Figure 2.16.

From these results it is obvious that the initial corridor width has not much weight on the optimal result but rather on the rate of convergence. Further, the larger the initial corridor width, the smaller the number of iterations required to converge to the optimum solution.

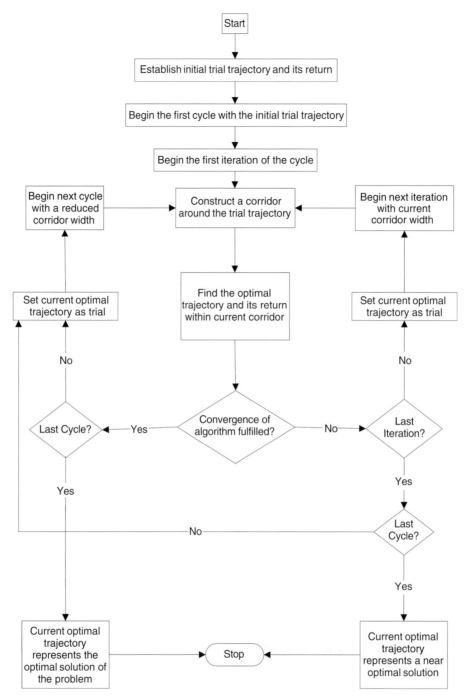

Figure 2.14 Incremental dynamic programming procedure

EFFECT OF INITIAL TRIAL TRAJECTORY

The model was run four times, each time the calculations starting from one of the trial trajectories 1, 2, 3, and 4 shown in Figure 2.17 with the same initial half corridor width. All four solutions converged to the same optimal trajectory as shown in the same figure. The rates of convergence to the optimal result are as shown in Figure 2.18. The optimum results obtained from the model for different initial trial trajectories are given in Table 2.9.

Table 2.9. *Effect of initial trial trajectory in IDP*

Trial trajectory number	Initial corridor width (10^6 m^3)	Number of iterations for convergence	Optimum annual energy generations (MWh)
1	40	20	1 465 174
2	40	46	1 465 174
3	40	19	1 465 174
4	40	27	1 465 174

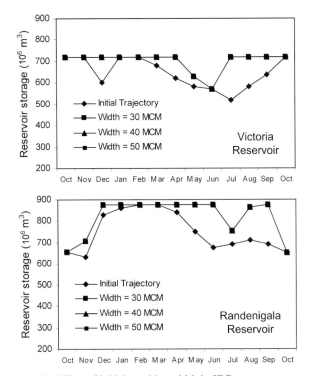

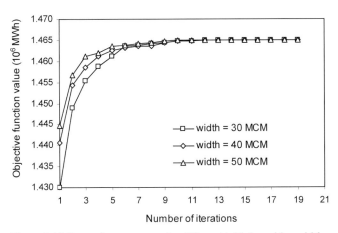

Figure 2.15 Effect of initial corridor width in IDP

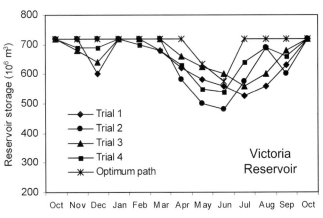

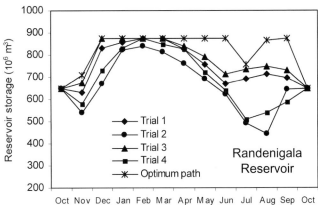

Figure 2.17 Effect of initial trial trajectory in IDP procedure

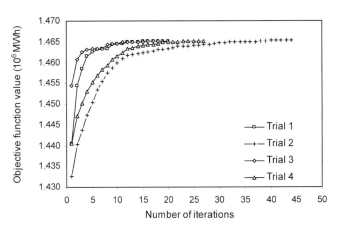

Figure 2.18 Rate of convergence for different initial trial trajectories in IDP procedure

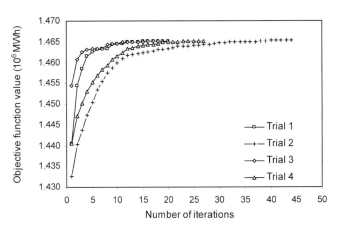

Figure 2.16 Rate of convergence for different initial corridor widths in IDP procedure

It is obvious from Figure 2.17 that the closer the initial trial trajectory is to the optimum result, the quicker is the convergence. As apparent from the above results, the convergence behavior of the model to the global optimum can be concluded as satisfactory.

The application of the IDP technique to the Mahaweli subsystem shows its suitability to derive optimum operational patterns simultaneously for two reservoirs in a system. It further presents the impact of different corridors as well as their widths on the rate of convergence to the optimum result and the optimum value in the IDP technique.

3 Stochastic dynamic programming in optimal reservoir operation

Uncertainty involved with water resources systems promotes the use of stochastic dynamic programming (SDP) in the derivation of optimum operation policies for reservoirs. The SDP procedure derives the optimal, expectation oriented, long-term operational strategy for reservoirs. A general description of the SDP technique is given in Section 1.6.

3.1 SDP IN OPTIMAL RESERVOIR OPERATION: SINGLE RESERVOIR

This section provides an SDP based model applicable in the operation of a single reservoir as shown in Figure 3.1 (He *et al.*, 1995).

OBJECTIVE FUNCTION

The objective is to maximize the expected annual energy generation from the reservoir:

$$\text{OF} = \text{Maximize } \xi \left\{ \sum_{j=1}^{T} \text{EP}_j \right\}, \tag{3.1}$$

where

EP_j = energy generation by power plant at period j (MWh)
  $= 9.81 \times \eta \times Q_j \times (\text{EL}_j - \text{TWL})/10^6, \quad j = 1, 2, \ldots, T,$
EL_i = average water surface elevation of reservoir during period j (m),
Q_j = release from reservoir during period j (m^3/s),
S_j = storage in reservoir at beginning of period j (10^6 m^3),
TWL_i = normal tail water level of power plant (m),
T = number of periods within annual cycle = 12,
η = overall efficiency of power plant (0.75 was used), and
ξ = denotes expectation.

STAGES, STATE, AND DECISION VARIABLES

The state of the system is described by water available in the reservoir at the beginning of any time step and inflow level at the present time period. Consecutive time steps are identified as stages. The decision variable is storage volume at the end of the time period. The optimization is subject to constraints on reservoir storage and release.

STORAGE VOLUME CONSTRAINT

The storage of the reservoir during any stage must be within the limits of minimum and maximum live storage capacity:

$$S_{\min} \leq S_j \leq S_{\max}; \quad j = 1, 2, \ldots, T, \tag{3.2}$$

where

S_j = storage volume at beginning of period j (10^6 m^3),
$S_{\min}$ = allowable minimum storage volume (10^6 m^3), and
$S_{\max}$ = allowable maximum storage volume (10^6 m^3).

RELEASE CONSTRAINT

The release from the reservoir at any stage is subject to the constraints of maximum and minimum limits. The capacity of hydropower generators sets a maximum limit to reservoir release:

$$R_{\min} \leq R_j \leq R_{\max}; \quad j = 1, 2, \ldots, T, \tag{3.3}$$

where

R_j = reservoir release during period j (10^6 m^3),
$R_{\max}$ = maximum allowable release through turbines in period j (10^6 m^3), and
$R_{\min}$ = minimum release from reservoir during period j (10^6 m^3).

STATE TRANSFORMATION EQUATION

The state transformation equation based on the principle of continuity is as follows:

$$S_{j+1} = S_j + I_j - E_j - R_j - O_j, \tag{3.4}$$

where

E_j = evaporation from reservoir during period j (10^6 m^3),
I_j = inflow to reservoir during period j (10^6 m^3), and
O_j = spillage water during period j (10^6 m^3),
$O_j = \text{Max}[S_j + I_j - E_j - R_j - S_{\max}, 0].$
  Other variables are as defined before.

Table 3.1. *Operational performance of the Kariba Reservoir*

Expected SDP based annual energy	8679 GWh
Simulated mean annual energy	8502 GWh
Minimum annual energy	6157 GWh
Mean utilized storage volume as % of available reservoir capacity	58.7%

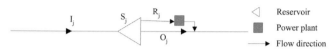

Figure 3.1 System configuration for SDP model: single reservoir

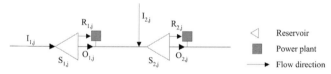

Figure 3.2 Graphical display of the indices used in the SDP model description

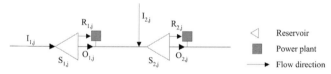

Figure 3.3 System configuration for SDP model: multiple-reservoir system

RECURSIVE EQUATION

The recursive equation for SDP optimization is the following:

$$F_j^n(k,p) = \max_l \left\{ B_{k,p,l,j} + \sum_q P_{p,q}^j \times F_{j+1}^{n-1}(l,q) \right\}, \quad (3.5)$$

where

k = storage state space values of reservoir at beginning of period j,

l = decision space values of storage state of reservoir at beginning of period $j+1$,

p = inflow state space values of inflow states during period j,

q = inflow state space values of inflow states during period $j+1$,

$F_j^n(k,p)$ = accumulated expected energy generation by optimal operation of reservoir over last n stages in GWh (when storage class at beginning of period j is k and inflow class during period j is p),

$B_{k,p,l,j}$ = energy generation when reservoir changes from state k to state l when inflow class is p in period j (GWh), and

$P_{p,q}^j$ = transition probabilities of inflows as defined by Eq. (3.6).

The transition probability $P_{p,q}^j$ is the probability that the inflow to the reservoir at period $j+1$ falls in state q given that at period j the streamflow to the reservoir was in state p. This can be expressed as

$$P_{p,q}^j = \text{prob}(I_{j+1} = q | I_j = p), \quad (3.6)$$

also

$$0 \leq P_{p,q}^j \leq 1.0, \quad \text{for all } p \text{ and } q; \qquad j = 1, 2, \ldots, 12, \quad (3.7)$$

$$\sum_q P_{p,q}^j = 1.0, \quad \text{for all } p; \qquad j = 1, 2, \ldots, 12, \quad (3.8)$$

where

I_j = inflow to reservoir during period j (10^6 m^3), $j = 1, 2, \ldots, T$. The absolute and monthly indices used to denote the stages of the recursive optimization process are displayed in Figure 3.2. The outline of the SDP procedure is displayed in Figure 1.4. Convergence criteria given in Section 1.6 are used in the model.

APPLICATION OF SDP TO THE KARIBA RESERVOIR

The model has been applied to the Kariba Reservoir considering monthly time steps. Varying inflow classes with equal occupancy frequencies and 33 storage classes with equal size were used in the study. It used inflows over the period from 1961 to 1984. Table 3.1 presents the results obtained from the model. In the simulation, the historical inflow time series have been used strictly relying on the SDP based operation policies.

3.2 SDP IN OPTIMAL RESERVOIR OPERATION: MULTIPLE-RESERVOIR SYSTEM

The applicability of the SDP technique for optimizing the operation of two-unit reservoir systems is presented based on an SDP model developed for the serially linked two-reservoir system displayed in Figure 3.3 (Kularathna, 1992).

OBJECTIVE FUNCTION

The objective is to maximize the expected annual energy generation from the system:

$$\text{OF} = \text{Maximize } \xi \left\{ \sum_{j=1}^T \left[\sum_{i=1}^2 \text{TEP}_{i,j} \right] \right\}, \quad (3.9)$$

where

$TEP_{i,j}$ = energy generation by power plant i at period j (MWh)

$\quad = 9.81 \times \eta \times Q_{i,j} \times (EL_{i,j} - DWL_{i,j})/10^6; \quad i = 1, 2;$
$\quad j = 1, 2, \ldots, T,$

$DWL_{i,j}$ = average downstream water level of power plant i during period j (m)

$\quad = \begin{cases} \max[TWL_1, EL_{2j}]; & i = 1; j = 1, 2, \ldots, T, \\ TWL_2; & i = 2; j = 1, 2, \ldots, T, \end{cases}$

$EL_{i,j}$ = average water surface elevation of reservoir i during period j (m),

$Q_{i,j}$ = release from reservoir i during period j (m^3/s),

$S_{i,j}$ = storage in reservoir i at beginning of period j (10^6 m^3),

TWL_i = normal tail water level of power plant i (m),

T = number of periods within annual cycle = 12,

η = overall efficiency of power plant (0.75 was used), and

ξ = denotes expectation.

The optimization is subject to constraints on reservoir storage and release.

STORAGE CONSTRAINT

The storage of the reservoirs during any stage must be within the limits of minimum and maximum live storage capacity.

$$SMIN_{i,j} \leq S_{i,j} \leq SMAX_{i,j}; \quad i = 1, 2; j = 1, 2, \ldots, T, \tag{3.10}$$

where

$SMIN_{i,j}$ = minimum storage of reservoir i at beginning of period j (10^6 m^3), and

$SMAX_{i,j}$ = maximum storage of reservoir i at beginning of period j (10^6 m^3).

RELEASE CONSTRAINT

The releases from each reservoir are subject to the constraints of maximum and minimum limits. This is due to the maximum capacities of outlets and the compulsory releases, if any:

$$RMIN_{i,j} \leq R_{i,j} \leq RMAX_{i,j}; \quad i = 1, 2; j = 1, 2, \ldots, T, \tag{3.11}$$

where

$R_{i,j}$ = release from reservoir i during period j (10^6 m^3),

$RMIN_{i,j}$ = minimum release from reservoir i during period j (10^6 m^3), and

$RMAX_{i,j}$ = maximum release from reservoir i during period j (10^6 m^3).

STATE TRANSFORMATION EQUATIONS

State transformation equations according to the principle of continuity are presented in the following.

For the upstream reservoir:

$$S_{1,j+1} = S_{1,j} + I_{1,j} - E_{1,j} - R_{1,j} - O_{1,j}; \quad j = 1, 2, \ldots, T. \tag{3.12}$$

For the downstream reservoir, since the releases and spills of the upstream reservoir become additional inflows:

$$S_{2,j+1} = S_{2,j} + I_{2,j} - E_{2,j} - R_{2,j} + R_{1,j} + O_{1,j} - O_{2,j}; \quad j = 1, 2, \ldots, T. \tag{3.13}$$

For both reservoirs:

$$O_{i,j} = R_{i,j} - RMAX_{i,j}, \text{ if } R_{i,j} \geq RMAX_{i,j} \text{ and } S_{i,j} \leq SMAX_{i,j}; \quad i = 1, 2; j = 1, 2, \ldots, T; \tag{3.14}$$

$$R_{i,j} = RMAX_{i,j}, \quad \text{when } R_{i,j} \geq RMAX_{i,j}; \quad i = 1, 2; j = 1, 2, \ldots, T; \tag{3.15}$$

and

$$O_{i,j} = 0.0, \quad \text{when } R_{i,j} \leq RMAX_{i,j}; \quad i = 1, 2; j = 1, 2, \ldots, T, \quad S_{i,j+1} \leq SMAX_{i,j+1} \tag{3.16}$$

$$S_{i,j+1} = S_{i,j}; \quad i = 1, 2; j = T, \tag{3.17}$$

where

$E_{i,j}$ = losses (principally evaporation) from reservoir i during period j (10^6 m^3),

$I_{i,j}$ = incremental inflow to reservoir i during period j (10^6 m^3),

$O_{i,j}$ = spill from reservoir i during period j (10^6 m^3), and

$R_{i,j}$ = release from reservoir i during period j (10^6 m^3).

Other variables are as defined before.

RECURSIVE EQUATION

The recursive equation for SDP optimization is the following:

$$F_j^n(k, p) = \max_l \left\{ B_{k,p,l,j} + \sum_q JP_{p,q}^j \times F_{j+1}^{n-1}(l, q) \right\}, \tag{3.18}$$

where

k = storage state space consisting of representative values of joint storage states of reservoirs at beginning of period j,

l = decision space consisting of representative values of joint storage states of reservoirs at beginning of period $j + 1$,

p = inflow state space consisting of representative values of joint inflow states during period j,

q = inflow state space consisting of representative values of joint inflow states during period $j + 1$,

$F_j^n(k, p)$ = accumulated expected energy generation by optimal operation of system over last n stages in GWh (when storage class at beginning of period j is k and inflow class during period j is p),

$B_{k,p,l,j}$ = energy generation when system changes from state k (reservoir 1 and reservoir 2 at states k_1 and k_2) to state l (reservoir 1 and reservoir 2 at states l_1 and l_2) when inflow class is p (p_1 to reservoir 1 and p_2 to reservoir 2) in period j (GWh), and

$JP^j_{p,q}$ = joint transition probabilities of inflows as defined by Eq. (3.19).

The joint transition probability $JP^j_{p,q}$ is the probability that the inflows to reservoir 1 and reservoir 2 at period $j+1$ fall in states q_1 and q_2 (represented by state vector q) given that at period j the streamflows to reservoirs 1 and 2 were in states p_1 and p_2 (represented by state vector p) respectively. This can be expressed as

$$JP^j_{p,q} = \text{prob}(I_{1,j+1} = q_1, I_{2,j+1} = q_2 | I_{1,j} = p_1, I_{2,j} = p_2),$$
(3.19)

also

$$0 \leq JP^j_{p,q} \leq 1.0, \quad \text{for all } p \text{ and } q; \qquad j = 1, 2, \ldots, 12,$$
(3.20)

$$\sum_q JP^j_{p,q} = 1.0, \quad \text{for all } p; \qquad j = 1, 2, \ldots, 12,$$
(3.21)

where

$I_{i,j}$ = inflow to reservoir i during period j (10^6 m^3), $i = 1, 2$; $j = 1, 2, \ldots, T$.

The absolute and monthly indices used to denote the stages of the recursive optimization process are displayed in Figure 3.2. The outline of the SDP procedure is displayed in Figure 3.4. Convergence criteria given in Section 1.6 are used in the model.

3.2.1 Application of SDP to the Mahaweli water resources system

Applicability of the model is shown based on the Victoria–Randenigala–Rantembe reservoir subsystem of the Mahaweli Development Scheme given in Figure 2.11. The Rantembe Reservoir, due to its negligible storage capacity, is treated as a run-of-the-river power plant (Nandalal, 1986). The objective function is to maximize the expected energy generation. The analysis is based on historical (37-year-long) monthly streamflow data at each reservoir and at the Minipe diversion. No irrigation demand constraints were considered in order to permit a comparison with the IDP based deterministic optimum. In the case of the deterministic optimum solution, it was found that a feasible solution does not exist when the available demand series are considered as constraints. Therefore, the SDP based optimization also was performed without demand constraints.

The operation policy designated for a reservoir by the model is a set of rules specifying the storage level at the beginning of the next month for each combination of storage levels at the beginning of the current month and the inflow during the current month. This optimization model produces an output consisting of 12 operation policy tables for the 12 months of the year. As an example, the operation policy table obtained for a month using four inflow classes and seven storage classes for each reservoir is displayed in Table 3.2. The numerical values used to identify the different inflow and storage levels are presented in Tables 3.3 and 3.4, respectively.

Once operational strategies have been defined, simulations are carried out to assess and to incorporate the effects of the reservoir's performance into the operation of the system as a whole. The simulations are carried out over the total historical record of inflow using different SDP based operation policies derived using different state discretization levels. The performance indicators used to assess the simulated operation are the following:

(a) average annual energy generation,
(b) annual firm energy,
(c) average annual water shortage,
(d) probability of failure months (the probability that the irrigation demand cannot be satisfied as a result of a reservoir level being lower than or equal to the minimum operation level).

The simulation results are summarized in Table 3.5. It also includes results of the deterministic optimum operation. The tabulated computer time is for the IBM 3083 mainframe computer at the Asian Institute of Technology in Thailand in 1986. These time requirements are clearly no longer representative. However, they still reflect the ratio in time requirement for IDP and SDP including the role of refining discretization.

Table 3.5 indicates that the computational time of an SDP model increases polynomially (for a fixed state space dimension) with the increase of state discretization levels ($NI_1 \times NI_2 \times NS_1 \times NS_2$ in Table 3.5). Although the memory requirements increase, they are well within the maximum memory limits of most modern personal computers. An improvement of the objective achievement can be noted when refining the storage discretizations. However, as demonstrated by Bogardi et al. (1988), the performance with respect to the refinement of state discretizations will eventually have a diminishing improvement as explicitly indicated by the increase of simulated annual energy generation. The SDP based policy No. 4 (Table 3.5) is observed to be the best policy for this water resources system when considering the annual firm energy generation. In terms of energy

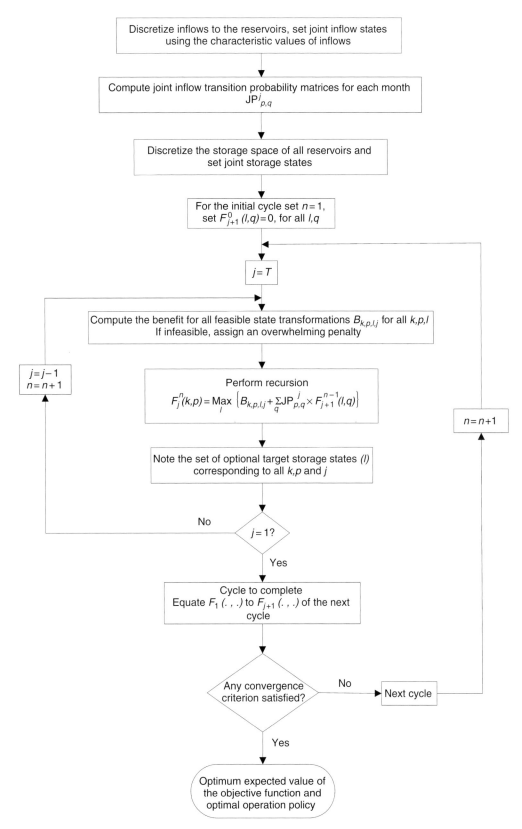

Figure 3.4 SDP Flow diagram for two-reservoir case

Table 3.2. *SDP based operation policy for the Victoria and Randenigala Reservoirs for the month of October*

Initial storage class	Inflow class of the current month															
	1	2	3	4	5	6	7	8	9	10	11	12	13	14	15	16
1	2	2	2	2	8	9	9	9	16	16	16	16	22	23	23	23
2	3	3	3	3	9	10	10	10	17	17	17	17	23	24	24	24
3	4	4	4	4	10	11	11	11	18	18	18	18	24	25	25	25
4	5	5	5	5	11	12	12	12	19	19	19	19	25	26	26	26
5	6	6	6	6	12	13	13	13	20	20	20	20	26	27	27	27
6	7	7	7	7	13	14	14	14	21	21	21	21	27	28	28	28
7	7	7	7	7	14	14	14	14	21	21	21	21	28	28	28	28
8	9	9	9	9	15	16	16	16	23	23	23	23	24	30	30	30
9	10	10	10	10	16	17	17	17	24	24	24	24	25	31	31	31
10	11	11	11	11	17	18	18	18	25	25	25	25	26	32	32	32
11	12	12	12	12	18	19	19	19	26	26	26	26	27	33	33	33
12	13	13	13	13	19	20	20	20	27	27	27	27	33	34	34	34
13	14	14	14	14	20	21	21	21	28	28	28	28	34	34	34	34
14	14	14	14	14	21	21	21	21	28	28	28	28	34	34	34	34
15	16	16	16	16	22	23	23	23	30	30	30	30	31	31	31	31
16	17	17	17	17	23	24	24	24	31	31	31	31	32	32	32	32
17	18	18	18	18	24	25	254	25	32	32	32	32	33	33	33	33
18	19	19	19	19	25	26	26	26	33	33	33	33	34	34	34	34
19	20	20	20	20	26	27	27	27	34	34	34	34	35	35	35	35
20	21	21	21	21	27	28	28	28	35	35	35	35	35	35	28	28
21	21	21	21	21	28	28	28	28	35	35	28	28	35	35	28	28
22	23	23	23	18	24	30	30	30	31	31	31	26	32	32	32	32
23	24	24	24	24	25	31	31	31	32	32	32	27	33	33	33	33
24	25	25	25	25	26	32	32	32	33	33	33	28	34	34	34	34
25	26	26	26	26	32	33	33	33	34	34	34	34	35	35	35	35
26	27	27	27	27	33	34	34	34	35	35	35	35	35	35	35	35
27	28	28	28	28	34	35	35	35	35	35	35	35	35	35	35	35
28	28	28	28	28	35	35	35	35	35	35	28	28	35	35	35	35
29	30	30	30	25	31	37	37	37	32	32	32	33	39	39	39	39
30	31	31	31	26	32	38	38	38	33	33	33	34	40	40	40	40
31	32	32	32	27	33	39	39	39	34	34	34	35	41	41	41	41
32	33	33	33	28	34	40	40	40	35	35	35	35	41	41	41	41
33	34	34	34	34	35	41	41	41	35	35	35	35	41	41	41	41
34	35	35	35	35	41	42	42	42	35	35	35	35	41	41	41	41
35	35	35	35	35	42	42	42	42	35	35	35	35	41	41	42	42
36	37	37	37	32	38	38	38	38	39	39	39	40	46	46	46	46
37	38	38	38	33	39	39	39	39	40	40	40	41	47	47	47	47
38	39	39	39	34	40	40	40	40	41	41	41	42	48	48	48	48
39	40	40	34	35	41	41	41	41	42	42	42	42	47	48	48	48
40	35	35	35	35	42	42	42	42	42	42	42	42	47	48	48	48
41	42	42	42	42	42	42	42	42	42	42	42	42	48	48	48	48
42	42	42	35	35	42	42	42	42	42	42	42	42	48	48	48	49
43	38	38	38	39	39	39	39	39	46	46	46	47	47	47	47	48
44	39	39	39	40	40	40	40	40	47	47	47	48	48	48	48	48
45	40	40	34	35	41	41	41	41	48	48	48	48	48	48	48	48
46	35	35	35	42	42	42	42	42	48	48	48	48	48	48	48	48
47	42	42	42	42	42	42	42	42	48	48	48	48	48	48	48	48
48	42	42	42	42	42	42	42	42	48	48	48	48	48	49	49	49
49	42	42	35	35	42	42	42	42	48	48	48	49	49	49	49	49

Final storage class target

Table 3.3. *Inflow class discretization of the operation policy of Table 3.2*

Inflow class		Oct.	Nov.	Dec.	Jan.	Feb.	Mar.	Apr.	May	Jun.	Jul.	Aug.	Sep.
1	V	93.5	155.1	146.2	67.2	34.7	25.6	41.2	47.0	87.5	79.3	102.7	99.3
	R	24.9	32.0	76.8	61.0	24.7	13.5	12.7	14.0	10.6	8.3	10.7	8.6
2	V	93.5	155.1	146.2	67.2	34.7	25.6	41.2	47.0	87.5	79.3	102.7	99.3
	R	43.1	68.2	161.9	118.0	89.9	36.3	26.8	33.9	24.2	21.8	22.5	23.7
3	V	93.5	155.1	146.2	67.2	34.7	25.6	41.2	47.0	87.5	79.3	102.7	99.3
	R	65.3	98.2	295.2	177.6	157.5	71.1	43.0	66.1	39.8	38.0	34.5	43.1
4	V	93.5	155.1	146.2	67.2	34.7	25.6	41.2	47.0	87.5	79.3	102.7	99.3
	R	89.7	141.2	447.0	255.7	248.0	106.1	59.7	96.3	67.6	54.9	58.2	57.1
5	V	176.2	279.2	381.8	200.5	106.2	48.4	80.9	166.7	280.8	223.4	236.6	261.2
	R	24.9	32.0	76.8	61.0	24.7	13.5	12.7	14.0	10.6	8.3	10.7	8.6
6	V	176.2	279.2	381.8	200.5	106.2	48.4	80.9	166.7	280.8	223.4	236.6	261.2
	R	43.1	68.2	161.9	118.0	89.9	36.3	26.8	33.9	24.2	21.8	22.5	23.7
7	V	176.2	279.2	381.8	200.5	106.2	48.4	80.9	166.7	280.8	223.4	236.6	261.2
	R	65.3	98.2	295.2	177.6	157.5	71.1	43.0	66.1	39.8	38.0	34.5	43.1
8	V	176.2	279.2	381.8	200.5	106.2	48.4	80.9	166.7	280.8	223.4	236.6	261.2
	R	89.7	141.2	447.0	255.7	248.0	106.1	59.7	96.3	67.6	54.9	58.2	57.1
9	V	304.5	429.7	710.2	313.0	169.9	75.6	131.0	260.5	440.1	312.1	365.7	422.4
	R	24.9	32.0	76.8	61.0	24.7	13.5	12.7	14.0	10.6	8.3	10.7	8.6
10	V	304.5	429.7	710.2	313.0	169.9	75.6	131.0	260.5	440.1	312.1	365.7	422.4
	R	43.1	68.2	161.9	118.0	89.9	36.3	26.8	33.9	24.2	21.8	22.5	23.7
11	V	304.5	429.7	710.2	313.0	169.9	75.6	131.0	260.5	440.1	312.1	365.7	422.4
	R	65.3	98.2	295.2	177.6	157.5	71.1	43.0	66.1	39.8	38.0	34.5	43.1
12	V	304.5	429.7	710.2	313.0	169.9	75.6	131.0	260.5	440.1	312.1	365.7	422.4
	R	89.7	141.2	447.0	255.7	248.0	106.1	59.7	96.3	67.6	54.9	58.2	57.1
13	V	402.4	657.5	1113.9	485.1	287.2	110.9	195.7	376.9	723.9	500.3	542.6	687.3
	R	24.9	32.0	76.8	61.0	24.7	13.5	12.7	14.0	10.6	8.3	10.7	8.6
14	V	402.4	657.5	1113.9	485.1	287.2	110.9	195.7	376.9	723.9	500.3	542.6	687.3
	R	43.1	68.2	161.9	118.0	89.9	36.3	26.8	33.9	24.2	21.8	22.5	23.7
15	V	402.4	657.5	1113.9	485.1	287.2	110.9	195.7	376.9	723.9	500.3	542.6	687.3
	R	65.3	98.2	295.2	177.6	157.5	71.1	43.0	66.1	39.8	38.0	34.5	43.1
16	V	402.4	657.5	1113.9	485.1	287.2	110.9	195.7	376.9	723.9	500.3	542.6	687.3
	R	89.7	141.2	447.0	255.7	248.0	106.1	59.7	96.3	67.6	54.9	58.2	57.1

V and R indicate the inflows of the Victoria and Randenigala Reservoirs respectively

Table 3.4. *Storage classes of the operation policy of Table 3.2*

Class	Vic.	Rand.	Class	Vic.	Rand.	Class	Vic.	Rand.	Class	Vic.	Rand.
1	34.0	295.0	13	148.0	778.0	25	377.0	585.0	37	605.0	390.0
2	34.0	390.0	14	148.0	875.0	26	377.0	682.0	38	605.0	488.8
3	34.0	488.8	15	262.0	295.0	27	377.0	778.0	39	605.0	585.0
4	34.0	585.0	16	262.0	390.0	28	377.0	875.0	40	605.0	682.0
5	34.0	682.0	17	262.0	488.8	29	490.0	295.0	41	605.0	778.0
6	34.0	778.0	18	262.0	585.0	30	490.0	390.0	42	605.0	875.0
7	34.0	875.0	19	262.0	682.0	31	490.0	488.8	43	720.0	295.0
8	148.0	295.0	20	262.0	778.0	32	490.0	585.0	44	720.0	390.0
9	148.0	390.0	21	262.0	875.0	33	490.0	682.0	45	720.0	488.8
10	148.0	488.8	22	377.0	295.0	34	490.0	778.0	46	720.0	585.0
11	148.0	585.0	23	377.0	390.0	35	490.0	875.0	47	720.0	682.0
12	148.0	682.0	24	377.0	488.8	36	605.0	295.0	48	720.0	778.0
									49	720.0	875.0

Vic. and Rand. indicate the storage volumes of the Victoria and Randenigala Reservoirs respectively

Table 3.5. *Simulation results of the Victoria–Randenigala–Rantembe reservoir subsystem according to SDP based policies*

Policy No.	Number of state discretization	Average annual energy (GWh)	Annual firm energy (GWh)	Average annual shortage at Minipe (10^6 m^3)	Probability of failure months[a] (%)	Size of DP program (bytes)	CPU time (s)
1	$4 \times 4 \times 4 \times 4 = 256$[b]	1265.9	150.8	93.1	5.4	112 544	38
2	$4 \times 4 \times 5 \times 5 = 400$	1274.3	153.1	85.5	5.4	172 660	94
3	$4 \times 4 \times 6 \times 6 = 576$	1284.0	123.1	84.1	5.4	260 108	195
4	$4 \times 4 \times 7 \times 7 = 784$	1283.0	164.3	85.1	5.4	383 396	365
Deterministic optimum		1427.5	67.8	552.0	38.1	319 328	274

[a] Failure to satisfy the irrigation water demands
[b] $NI_1 \times NI_2 \times NS_1 \times NS_2$
NI_i and NS_i are respectively the number of inflow discretizations and the number of storage discretizations for the *i*th reservoir

generation and average water shortage this policy is negligibly inferior when compared to policy No. 3. The underachievement with respect to energy generation and the average water shortage are 0.07% and 1.18% respectively. However, the overachievement in terms of firm energy (33%) confirms the acceptance of policy No. 4 as the best policy. A comparison of this policy with the deterministic optimum reveals that it has achieved 89.9% of the deterministic optimum energy generation.

3.3 SOME ALGORITHMIC ASPECTS OF STOCHASTIC DYNAMIC PROGRAMMING

The previous sections reveal the great potential of SDP in optimal reservoir operation. SDP, which can handle non-convex and nonlinear discrete variables, generates an operation policy comprising storage targets, release decisions for all the possible reservoir storages, and inflow states in each period rather than a mere single schedule of reservoir releases. It is a flexible model that could be adjusted easily to various problem environments. Due to its inherent merits, SDP has been well received as a long-term reservoir optimization model.

However, many applications of SDP indicate that certain algorithmic aspects of it have to be studied further to facilitate the application of the SDP model to real-world reservoir operational problems.

He *et al.* (1995) studied algorithmic aspects of SDP based on its application to several real-world reservoir operational problems. The study focused on the following aspects of the SDP model: (a) the Markov inflow transition probability matrix and its role in SDP models; (b) the influence of different decision variables and inflow state variables on

performance of the SDP model; and (c) the suitability of the different inflow serial correlation assumptions.

3.3.1 Markov inflow transition probabilities

Application of the SDP technique for the optimization of reservoir operations is based on the idea that the policies will converge to a "steady-state" policy after several iterations of the recursive relation. The steady-state policy achieved in this way will be the global optimum. In several applications of the SDP technique in reservoir operation optimizations (e.g., Nandalal, 1986; Budhakooncharoen, 1986; Kularathna, 1992, etc.), one convergence criterion given in Section 1.6, that is the stabilization of the expected annual increment of objective function value, could not be realized. In those studies only the stabilization of the operation policy after a few iteration cycles was reported to have been achieved.

PROBLEMS IN SDP CONVERGENCE BEHAVIOR
In the SDP model, the inflow process, *I*, is usually assumed to be a "Markov process" (or "Markov chain"). In general, a Markov process describes only one-step dependence, called a first order process, or exhibits lag-one serial correlation (Markov assumption). An SDP model is the application of the "principle of optimality" of dynamic programming (Bellman, 1957) to the Markov sequential decision process.

As an example, consider the version of the SDP model presented in Section 3.1. In that model, time period is defined as the stage. Storage volume at the beginning of the time period and inflow level at the present time period are state variables. Storage volume at the end of the time period is defined as the decision variable. Markov transition probabilities of the inflow (from the present time period to the subsequent time period) are incorporated into the recursive relation to derive optimal values.

The optimization process starts with a set of initial values $F_T^1(k, p)$. Due to the characteristics of the Markov sequential decision process, after a large number of iterations of the recursive relation (Eq. 3.5) the "steady state" for each period in successive years will finally be reached. It is independent of the initial state.

There are two criteria marking convergence of the steady state: (i) stabilization of the policy; and (ii) stabilization of the expected annual increment of the objective values. The interpretation of the two criteria is presented in Section 1.6. However, experimental evidence shows that often the second criterion of convergence cannot be achieved.

REASONS FOR THE VIOLATION OF SDP CONVERGENCE CRITERIA

The behavior of the policy convergence after many iterations is the resulting performance of the Markov transition probability matrix incorporated in the recursive relation that converges to its steady-state probabilities. Howard (1960) proved that the policies would converge to the global optimum if the associated Markov transition probability matrices were "ergodic." Stationarity means that the probability distribution of the inflow process is not changing over the time cycle (Loucks et al., 1981). This is the condition necessary to ensure that the policies will become stable after a certain number of iterations (the first convergence criterion).

A Markov chain is said to be ergodic if all the members in the chain form a single recurrent chain. Immaterial of the starting point, the process would end making jumps among all the members in the chain. In other words, ergodicity implies that the final state of the system is independent of the initial state (Howard, 1960). This condition ensures that the stable policies will be the global optimum.

In practice, the transition probabilities are usually estimated from observed inflow records. This is done by counting the number of times the observed data transit from state I_{j-1} in period $j-1$ to I_j in period j. This simple method is suitable when the number of inflow classes is small. However, it has the drawback of limiting the accuracy of the SDP model.

When the number of inflow classes is larger than 3 or 4, a difficulty arises. For example, if the inflows of subsequent periods are discretized into 10 classes, the number of elements to be estimated in a matrix during a period is $10 \times 10 = 100$. In practice, historical reservoir inflow time series are seldom longer than 50 years. For developing countries, a 30-year record is considered as a long record. A large number of elements will remain void if 30 years of monthly inflow data are used to estimate 12 matrices (each with 100 elements) in a

year. This would cause real danger of losing the ergodicity of the matrices. These zeros are "artificial" in the sense that they are due to the small size of the sample.

In the SDP model, the problem is more aggravated as there are T (e.g., 12 with a monthly time period) Markov inflow transition probability matrices in one year cycle. The optimal policies are produced after many cycles (years) of iterative calculation. Therefore, it is not easy to judge whether the ergodicity requirement is satisfied by looking at the combination of T matrices. For example, some vectors that do not communicate in the matrix of transitions from October to November may communicate in the matrix of November to December.

One definite sign of the combination matrices having ergodicity is that the second criterion for steady-state policy, that is a constant value of annual increment of objective function, can be achieved.

The importance of the ergodicity property of Markov chains has been disregarded in their application in the SDP model. The problem stated at the beginning of this section, that is the difficulty in achieving the second convergence criterion in many SDP applications, can be explained by the above discussion. For those cases, stable policies can be reached after a few iteration cycles while the annual increment of the objective function value converges to more than one constant (instead of one). The failure to converge to a single constant in the annual increment of objective value is due to the violation of ergodicity of the Markov inflow processes. At this time, although the first convergence criterion (i.e., stable policies) is obtained, the set of stable policies is separated into more than one group, which have no communication among each other. Each group obtains its optimum with respect to the initial state when the reservoir operation starts.

The cause of ergodicity violation can be traced back to the large number of zero-elements in the estimated transition probability matrices. Therefore, an important point obtained from this analysis is that, in applying SDP model, the number of zero-elements in the reservoir inflow transition probability matrices should be kept within a limit to guarantee that the derived policies will be a global optimum.

Therefore, an approach to satisfy the ergodicity requirement is needed while keeping the computing effort requirements at a reasonable level. The method of deriving inflow transition probabilities by distribution fitting involves a considerably large amount of computing effort. This suggests the necessity of finding an alternative method to smooth out the zero-elements in the matrices derived from the simple tabulating method. Therefore, the elimination of zero-elements while maintaining the performance of the derived optimal operation policies is of interest.

Table 3.6. *SDP model setups for the Mahaweli and Kariba reservoir systems*

	Mahaweli system: Victoria and Randenigala Reservoirs	Kariba Reservoir
Objective	Maximize expected annual energy generation	
Constraint	Irrigation demand	—
Inflow discretizations	Equal size intervals 4×4 $= 16$ combinations	Equal occupancy varying number of classes (from 2 to 8)
Storage discretizations	$7 \times 7 = 49$ combinations	33 classes
Time step length	One month	One month

SENSITIVITY ANALYSIS OF MARKOV INFLOW TRANSITION PROBABILITIES

The impact of transition probability matrices on reservoir operational performance is presented based on the Kariba Reservoir and the Mahaweli reservoir system. Several hypothetical transition probability matrices reflecting different flow regimes are used.

The sensitivity is carried out for both systems based on the following steps:

(a) set up SDP models,
(b) create several sets of extremely different inflow transition probability matrices and incorporate them into the SDP model to derive several sets of operation policies, and
(c) simulate with the historical inflow time series according to the derived sets of policies and compare the resulting performance.

Table 3.6 presents the SDP models used to derive operation policies for the two reservoir systems. The versions of SDP model with the recursive relations described in Eq. (3.5) and Eq. (3.18) were used for the Kariba Reservoir and the Mahaweli reservoirs, respectively.

The observed original transition probabilities (ORG) and some modified forms, namely, modified transition probabilities (MDF), average transition probabilities (AVG) and modified average transition probabilities (AVM) have been adopted in the SDP models to derive respective optimal operation policies.

Original transition probabilities are derived from the historic inflow series. Modified transition probabilities are obtained from the ORG version by overemphasizing the maximum probability (or probabilities) occurring in every line. Average transition probabilities are assumed to characterize a hypothetical uniform frequency distribution of inflow class transitions. Modified average transition probabilities combine the principles of MDF and AVG. Zero-elements are kept zero at the beginning and at the end of each row. Internal sequences of three or more zero-elements remain as

Table 3.7. *Example of modifications of the Markov inflow transition probabilities of the Kariba Reservoir*

ORG	0.500	0.250	0.250	0.000	0.000	0.000	0.000	0.000
	0.000	0.000	0.000	0.333	0.000	0.333	0.000	0.333
	0.000	0.000	0.000	0.000	0.000	0.000	0.333	0.670
	...	...						
MDF	1.000	0.000	0.000	0.000	0.000	0.000	0.000	0.000
	0.000	0.000	0.000	0.333	0.000	0.333	0.000	0.333
	0.000	0.000	0.000	0.000	0.000	0.000	0.000	1.000
	...	...						
AVG	0.125	0.125	0.125	0.125	0.125	0.125	0.125	0.125
	0.125	0.125	0.125	0.125	0.125	0.125	0.125	0.125
	0.125	0.125	0.125	0.125	0.125	0.125	0.125	0.125
	...	...						
AVM	0.333	0.333	0.333	0.000	0.000	0.000	0.000	0.000
	0.000	0.000	0.000	0.200	0.200	0.200	0.200	0.200
	0.000	0.000	0.000	0.000	0.000	0.000	0.500	0.500
	...	...						

such, while a uniform frequency distribution is assumed row-wise over the nonzero and imbedded single or double zero-elements of the ORG matrices. Table 3.7 shows a few example lines of these inflow transition probabilities calculated for the Kariba Reservoir.

In the subsequent simulation, the historical inflow time series are used "strictly" relying on the SDP based operation policies obtained according to the different sets of transitional probabilities. The optimum operation policies are determined using the expected system performance based on discrete storage and inflow states. Therefore, it is possible that in some periods the actual releases resulting from the continuity equation will be out of their feasible range (e.g., release less than 0 or larger than the downstream channel capacity). In such instances, corrections (i.e., over-ruling the SDP optimum operation policy) are required in the simulation model. The releases are made equal to the nearest feasible value and the

Table 3.8. *Operational performance of the Kariba Reservoir*

	ORG	MDF	AVG	AVM
Indices referring to energy output				
(1) Expected annual energy (GWh)	8679	9164	9291	8967
(2) Simulated mean annual energy (GWh)	8502	8355	8494	8478
(3) Standard deviation of (2) (GWh)	847	1226	914	996
(4) Minimum annual energy (GWh)	6157	4606	5383	5358
Indices referring to reservoir				
(5) Mean utilized storage volume as % of available reservoir capacity	58.7%	57.5%	44.4%	50.6%
(6) Standard deviation of (5)	26.6%	26.9%	24.2%	26.0%
(7) Minimum storage drawdown as % of available reservoir capacity	0.0%	0.0%	0.0%	0.0%

consequent final storage is defined by the continuity equation. The final storage (or the closest discrete value) obtained is used as the initial storage for next time step.

The operation performances of the Kariba and Mahaweli systems are shown in Table 3.8 and Table 3.9, respectively. The expected annual energy outputs shown are obtained from the SDP optimization. They are the expected gains for the different sets of inflow transition probabilities considered and do not represent the real gain of the reservoir operation. The simulated mean annual energy outputs are obtained from the real-time operational model for the given time period. Those results are based on the operation policies derived from different sets of inflow transition probabilities. The performance based on "ORG policy" is observed to be better than the others. That operation has the largest average energy output, least standard deviation, and largest minimum energy output. The differences among mean annual energy outputs appear to be very limited (less than 2%). The standard deviations and the minimum energy output seem to be more sensitive for assumed changes in the inflow regime. However, the variation among the simulated performances of the reservoir system for different sets of assumed transition probability matrices is observed to be very much limited.

Table 3.8 also presents the mean utilized reservoir storage volumes, their standard deviations and the minimum storage drawdowns. The mean utilized reservoir storage volume related to "AVG policy" is much smaller than the value from "ORG policy". The "AVG policy" is derived based on the assumption that the reservoir inflow transition probabilities are uniform. This implies that there is a moderate (not low and not high) incoming inflow. If this assumption is valid, the reservoir storage capacity needed to regulate the over-year inflow is low. Therefore, decisions (the storage volumes at the end of each

time period) made by the SDP model lead to a smaller mean utilized reservoir storage volume.

Table 3.9 shows the performance of the Victoria and Randenigala Reservoirs in the Mahaweli system. In the SDP model, the objective of maximizing energy generation is subjected to the constraint of irrigation requirements. When both the inflow to the reservoir and the initial reservoir storage are very small it may not be possible to make a decision (reservoir storage at the end of the period) that falls into the feasible region. This occurs as the irrigation demand is introduced to the system as a constraint that has to be satisfied always. Zeros represent these cases in the operation policy tables. Since the optimization does not hold for the whole set of decisions in the annual cycle, the expected annual energy output is not obtained.

Table 3.9 presents the simulated mean annual energy outputs corresponding to the operation policies derived for different sets of inflow transition probabilities. The mean utilized reservoir storage volumes, their standard deviations, and the minimum drawdown of the reservoir storages are also given in the table. Except the indices referring to reservoir storage volume, most of the performance indices referring to the objective function (energy generation) display little variation among different sets of policies derived from different inflow series. These results are similar to those for the Kariba Reservoir.

The Mahaweli system serves two purposes: energy generation and irrigation supply. The irrigation requirement is set as a constraint in the SDP model. In Table 3.9, the performance indices referring to irrigation supply are also presented. They display little variation among different policies derived from different sets of inflow series.

These results show that for the given inflow and storage discretizations, a limited impact of the different transition probabilities can be detected as far as the objective functions

Table 3.9. *Operational performance of the Mahaweli system*

	ORG	MDF	AVG	AVM
Indices referring to energy output				
(1) Mean annual energy (GWh)	1390	1364	1342	1384
(2) Standard deviation of (1)	308	302	311	321
(3) Minimum annual energy (GWh)	841	838	774	730
Indices referring to reservoir				
(4) Mean utilized storage volumes as % of available reservoir capacity				
Victoria	80.1%	89.1%	42.4%	65.6%
Randenigala	91.2%	89.1%	86.5%	91.5%
(5) Standard deviation of (4)				
Victoria	6.6%	7.8%	11.3%	8.0%
Randenigala	10.2%	10.0%	11.8%	9.9%
(6) Minimum storage drawdown as % of available reservoir capacity				
Victoria	20.6%	36.5%	4.7%	20.6%
Randenigala	44.8%	44.8%	33.7%	44.8%
Indices referring to irrigation shortage				
(7) Time-based reliability[a]	86.2%	85.9%	84.6%	86.2%
(8) Quantity-based reliability[b]	95.9%	95.9%	95.5%	95.7%
(9) Repairability[c] (month)	1.57	1.64	1.64	1.47
(10) Vulnerability[d] (10^6 m^3)	60.5	61.6	62.0	59.6

[a] % of time steps with fulfilled irrigation demand
[b] % of accumulated irrigation demand met
[c] Average duration of an irrigation failure (shortage) event
[d] Average accumulated irrigation shortages per failure

are concerned. This fact implies that the inherent inaccuracy in estimating the transition probabilities is unlikely to have a considerable impact on the SDP based operational performance of reservoir systems. This "insensitivity" phenomenon may be interpreted from the following two aspects.

(a) The transition probability matrices with large variations may derive similar steady-state policies. As mentioned before, the SDP optimization process is an iteration of the Bellman recursive relation with incorporated transition probability matrices (Eq. 3.5). Each row of the transition probability matrix is associated with an expected objective value of a feasible region. The optimal decision for each state is selected from the whole set of feasible decisions for that state by comparing the expected value of those decisions. It is clear from the behavior of Markov chains that the influence of the initial transition probabilities is decreasing along the iterations. After many cycles the decisions are mainly weighted by the steady probabilities of the transitions.

However, the foregoing discussion does not imply that all the transition probability matrices with the same steady probabilities will always derive the same steady-state policy. Two transition probability matrices with large variation at the start

of the optimization process may be deformed to have less variation along the path of optimization iterations. Thus, they may derive similar optimal policies at the end.

(b) The derived operation policies, which vary to a certain extent, may satisfy the purpose of the reservoir system to similar standards.

Water reservoirs are expensive long-life investment projects. Once they have been built, they are often operated for decades. Therefore, when designing reservoirs the uncertainty of the future supplies, flows, qualities, costs, benefits, and so on has to be considered. While the forecast for the future conditions is never perfect, well-designed reservoirs need to be sufficiently flexible to permit their adaptation to a wide range of possible future conditions. Nowadays, many professionally designed reservoirs have to a certain extent built-in robustness for dealing with future uncertainties. Therefore, policies which vary to a certain extent from the optimal policies may not cause much worse performance of reservoir systems.

In the two case studies presented, the operation policies derived from transition probability matrices "AVG" vary from the policies derived from the matrices "ORG". This

Table 3.10. *Example of the smoothing method*

ORG	0.500	0.250	0.250	0.000	0.000	0.000	0.000	0.000
	0.000	0.000	0.000	0.333	0.000	0.333	0.000	0.333
	0.000	0.000	0.000	0.000	0.000	0.000	0.330	0.670
	...	...						
NEW	0.480	0.240	0.230	0.010	0.010	0.010	0.010	0.010
	0.010	0.010	0.010	0.313	0.010	0.323	0.010	0.313
	0.010	0.010	0.010	0.010	0.010	0.010	0.310	0.630
	...	...						

Table 3.11. *Simulated performance after smoothing*

	ORG	NEW
Indices referring to energy output		
(1) Expected annual energy (GWh)	8679	8693
(2) Simulated mean annual energy (GWh)	8502	8493
(3) Standard deviation of (2) (GWh)	847	847
(4) Minimum annual energy (GWh)	6157	6157
Indices referring to reservoir		
(5) Mean utilized storage volumes as % of available reservoir capacity	58.7%	58.8%
(6) Standard deviation of (5)	26.6%	26.7%
(7) Minimum storage drawdown as % of available reservoir capacity	0.0%	0.0%

can be detected from the indices referring to the reservoir storage level (policy is a set of decisions defining the reservoir storage at the end of each time period). However, the differences among the performance indices referring to the key concerns of the reservoir systems (energy output and irrigation supply) are limited.

A METHODOLOGY TO ELIMINATE ZEROS IN TRANSITION PROBABILITY MATRICES

As presented previously, the large number of zero-elements in the transition probability matrices due to the limited length of inflow record are the cause of the violation of the second convergence criterion of the SDP model. Therefore, when using the SDP model, it is safer to make sure that most of the elements in each row in the transition probability matrices are nonzero.

It can be easily seen that a transition probability matrix is ergodic (irreducible) if more than half the elements in each row are nonzero. This can be proved by the reduction to absurdity. Assume an $n \times n$ nonergodic matrix having more than $n/2$ nonzero elements in each row. This matrix can be divided into at least two groups of rows (each nonergodic matrix is reducible and can be reduced to at least two non-communicating groups of rows). These two groups of rows do not have nonzero elements at the same column. Therefore, these groups of rows form a matrix with more than n columns. This is in contradiction with the given fact that the matrix is an $n \times n$ matrix.

The results of the sensitivity analysis presented previously reveal that the inherent inaccuracy in estimating the transition probability matrices is unlikely to have a considerable impact on the SDP based operational performance of reservoir systems. Therefore, it is proposed that some zeros in the transition probability matrices may be easily smoothed out by a reasonably small value. The following example applied to the Kariba Reservoir illustrates this.

The method aims to make most of the elements in the transition probability matrices nonzero. When a row is empty, a

uniform frequency distribution is assumed row-wise. When more than half the elements of a row are nonzero, the row is kept unchanged. Otherwise, each zero in the row is replaced by 0.01 while the nonzeros in the row are accordingly reduced slightly to maintain the sum of the row to 1.0. Table 3.10 shows an example of how a given transition probability matrix is transformed into a new transition probability matrix by smoothing out the zeros according to the method.

The simulated reservoir operation performance based on the policy derived by incorporating the new transition probability matrices (after partially smoothing out zeros) in the SDP model is compared with that from the original transition probability matrices.

Table 3.11 shows the average annual performance indices from the simulated reservoir operation concerning both reservoir storage and energy output. The similarity in the performance indices for reservoir storage can be interpreted as the similarity of the operation policies derived from both "ORG" and "NEW" transition probability matrices. Their performance indices referring to the energy output are also similar.

3.3.2 State and decision variables

For any DP type of model, the careful choice of state and decision variables is crucial to the success of the model. There are two versions of stationary SDP models that have been applied in reservoir operation optimization. One is the model having release as the decision variable, with previous inflow and initial storage as state variables. The other is the model having final storage as the decision variable, with present inflow and initial storage as state variables. This section presents an insight into the roles of different decision variables and state variables in the SDP model and compares their relative performances during reservoir operation.

COMPARISON OF MODELS WITH DIFFERENT DECISION AND INFLOW STATE VARIABLES

This section compares the performance of several different SDP models. These models, which differ in the choice of inflow state variables and decision variables, are given below.

Model 1

State   Initial storage + Present inflow
Decision   Final storage

$$F_j^n(k,p) = \underset{l}{\text{Opt}} \left[B_{k,p,l,j} + \sum_q P_{p,q}^{j+1} \times F_{j+1}^{n-1}(l,q) \right] \qquad (3.22)$$

Model 2

State   Initial storage + Previous inflow
Decision   Final storage

$$F_j^n(k,m) = \underset{l}{\text{Opt}} \left[\sum_p P_{m,p}^j \left\{ B_{k,p,l,j} + F_{j+1}^{n-1}(l,p) \right\} \right] \qquad (3.23)$$

Model 3

State   Initial storage + Present inflow
Decision   Current release

$$F_j^n(k,p) = \underset{r}{\text{Opt}} \left[B_{k,p,r,j} + \sum_q P_{p,q}^{j+1} \times F_{j+1}^{n-1}(l,q) \right] \qquad (3.24)$$

Model 4

State   Initial storage + Previous inflow
Decision   Current release

$$F_j^n(k,m) = \underset{r}{\text{Opt}} \left[\sum_p P_{m,p}^j \left\{ B_{k,p,r,j} + F_{j+1}^{n-1}(l,p) \right\} \right] \qquad (3.25)$$

where

k = storage state space values of reservoir at beginning of period j,

l = decision space values of storage state of reservoir at beginning of period $j+1$,

m = inflow state space values of inflow states during period $j-1$,

p = inflow state space values of inflow states during period j,

q = inflow state space values of inflow states during period $j+1$,

r = decision space values of release state during period j,

Transition probabilities are defined as in Eq. (3.6).

Figure 3.5 shows the number of storage state space, inflow state space, and release state space values at different periods.

The four SDP formulations are compared by applying them to the Kariba Reservoir system. Initially the optimum operation

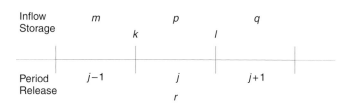

Figure 3.5 Number of inflow, storage, and release state space discretizations

policies were derived and then the system operations were simulated according to the derived policies. The comparison is made based on the performance indices obtained from the simulation.

EXPERIMENTS FOR COMPARISON OF THE MODELS

The objective of the models is to maximize the expected annual energy generation from the system. The stage of the models is the time period (one month). The optimization is subjected to physical constraints of the reservoir system (e.g., storage constraints, release constraints, etc.). The whole reservoir storage is discretized into 42 classes of equal size, each of 1579×10^6 m³. For the models in which policies are defined as optimal releases (Model 3 and Model 4), release levels up to twice the monthly release capacity of the turbines are to be optimized. They are discretized into 6 classes with equal size, each having an interval of about 1575×10^6 m³. The monthly inflows are discretized into varying numbers of classes (2 classes for August, September, October, and November; 3 classes for December and July; 4 classes for January and June; 6 classes for February and May; 8 classes for March and April) with equal occupancy frequencies. The median of each inflow class is defined as the representative value of that class. Uniform frequency distribution is assumed row-wise for the empty rows if they occur in the inflow transition probability matrices. Empty rows may occur due to the limited length of the observed historical inflow time series. For all the SDP based optimizations, both convergence criteria have been obtained.

In the subsequent simulations, the historical inflow time series were used "strictly" relying on the derived SDP based operation policies subjected to the physical constraints of the reservoir system. The following three experiments have been carried out based on the models.

EXPERIMENT 1

Derive optimal operation policies for the reservoir system using the four SDP models based on 24 years (1961–84) of historical inflow time series. Then simulate the performance

Table 3.12. *Multiple regression analysis of the Kariba Reservoir inflow (Budhakooncharoen, 1986)*
Multiple regression equation: $x_j = A_1 x_{j-1} + A_2 x_{j-2} + A_3 x_{j-3} + A_4$,
where R^2 is the determination coefficient

	Oct.	Nov.	Dec.	Jan.	Feb.	Mar.	Apr.	May	Jun.	Jul.	Aug.	Sep.
A_1	0.8773	0.3269	−0.8018	0.7772	0.8494	0.3063	1.0595	0.5854	0.6720	0.3446	0.4389	0.2915
A_2	0.2070	0.4326	3.5237	−0.5743	0.2684	0.4652	−0.2041	−0.0272	−0.1394	0.1413	−0.1464	0.0385
A_3	−0.2469	−0.1821	2.0885	−3.2355	0.6903	0.7587	−0.3044	0.0612	0.0088	0.0068	0.0727	0.0057
A_4	310.48	780.55	−2288.14	6967.94	899.77	1734.72	2845.72	2081.14	604.85	−166.32	602.65	752.66
R^2	0.382	0.192	0.227	0.353	0.441	0.431	0.843	0.841	0.896	0.970	0.924	0.637

Table 3.13. *Summary of the three computer experiments*

	Experiment 1	Experiment 2	Experiment 3
Inflow time series used to derive release policy	1961–84	1961–72	1961–72
Inflow time series used to simulate system operation	1973–84	1973–84	1973–84
Type of forecast available at beginning of each period (during simulation)	Perfect forecast	Perfect forecast	Imperfect forecast

of the reservoir system according to the derived operation policy sets using the last 12 years (1973–84) of historical inflow time series. Assume that the perfect forecast is available at the beginning of each time period.

EXPERIMENT 2

Derive optimal operation policies for the reservoir system using the four SDP models based on 12 years (1961–72) of historical inflow time series. Then simulate the performance of the reservoir system according to the derived operation policy sets using the last 12 years (1973–84) of historical inflow time series. Assume that the perfect forecast is available at the beginning of each time period.

EXPERIMENT 3

Derive optimal operation policies for the reservoir system using the four SDP models based on 12 years (1961–72) of historical inflow time series. Then simulate the performance of the reservoir system according to the derived operation policy sets using the last 12 years (1973–84) of historical inflow time series using the imperfect inflow forecast at the beginning of each time period. The inflows are forecast by using the regression equations derived by Budhakooncharoen (1986) for this case study system. The regression equations obtained from that study are shown in Table 3.12.

The three computer experiments are summarized in Table 3.13.

In Experiment 1, the part of the historical inflow series (1973–84) that has been employed for deriving the optimal policies is used to simulate the performance of the system. This enables the difference between system performance values obtained from the steady-state solution of SDP and that from simulation to be observed. SDP relies on a discrete representation of input states while simulation uses actual inflow series.

In Experiment 2, the first 12 years of the historical inflow series are employed to derive optimal policies. The second 12 years of the historical inflow series are used to simulate the performance of the system according to the derived policies. This corresponds to the situation that exists in reality if the system is operated based on a perfect inflow forecast.

The derivations of optimum operation policies in Experiment 2 and Experiment 3 are similar. But in the implementation of the derived policies in the operation simulations, the forecast inflow data are used in Experiment 3 while observed inflow data are used in Experiment 2. Nevertheless, when simulating the operation, the inflows used in the continuity equation are still the actual inflows. In Experiment 3, the models whose derived policies can be implemented with known previous inflow (Model 2 and Model 4) will perform the same way as in Experiment 2. Variation from Experiment 2 occurs only in those models whose derived policies have to be implemented using inflow forecast data.

ANALYSIS OF THE RESULTS AND DISCUSSION
Each experiment derives one SDP based optimal policy containing 12 tables (for 12 months). Out of the large number of policies derived, the policy tables for the month of May from Experiment 2 are presented in Table 3.14 for all four models as an example. Table 3.14a, Table 3.14b, Table 3.14c, and Table 3.14d refer to the policies from Model 1, Model 2, Model 3, and Model 4, respectively. Note that the policies from Experiment 3 are the same as those from Experiment 2. The values in Table 3.14a and Table 3.14b are the targeted storage classes in the reservoir at the end of the month. The values in Table 3.14c and Table 3.14d are the targeted release classes during the month.

The simulated performance is presented based on the following three aspects: (a) average reservoir storage; (b) average release through turbine; and (c) average energy generation. The energy generation, being the objective of the optimization, is the most important performance index. The release through turbine is directly proportional to the energy generation. Reservoir storage shows the behavior of the reservoir very clearly.

Table 3.15 presents the average annual performance indices regarding the reservoir storage, the release through turbines, and the energy generation from Experiment 1. It presents simulated mean, standard deviation, and minimum value for each of the three performance indices. For energy generation the expected annual gain obtained from the SDP based optimization is also included. To make the comparison easy, values corresponding to reservoir storage, turbine release, and energy generation are presented as percentages of the reservoir capacity, the average annual inflow, and the generation capacity, respectively.

Tables 3.16 and 3.17 present the average annual performance indices from Experiment 2 and Experiment 3, respectively.

In Table 3.14 the particular structure of the policy tables for models with release as the decision variable (Table 3.14c and Table 3.14d) attracts attention. It is noticeable that most of the decisions are Class 1, 3212×10^6 m^3, which is close to the capacity of the turbines (3937×10^6 m^3). The decision changes to the value of one class larger (4818×10^6 m^3) only if both initial storage and inflow are very large, or to the value of one class smaller (1606×10^6 m^3) only if both initial storage and inflow are very small. This structure of the policy table can be explained as follows. The objective of the optimization is to generate as much energy as possible. The maximum energy that a hydropower plant can generate is the capacity of its generators. The amount of energy generation is proportional to the release through the penstocks. Corresponding to the capacity of the generators there is an imaginary capacity

release (R_c). Therefore, optimal decisions can be seen as attempts to approach R_c.

In contrast to Table 3.14c and Table 3.14d, models with storage as the decision variable (Table 3.14a and Table 3.14b) have policy tables that show large variations in the decisions. The above-mentioned feature of "stability" is achieved for the policy with release as decision variable only when release is the direct target of optimization (e.g., to satisfy downstream water requirements or to satisfy energy requirements, etc.). The feature of "stability" would occur in the policy with storage as the decision variable if storage were the direct target of optimization. For example, if water level in a reservoir is important, the objective of optimization could be minimizing the deviation from a target storage level at each time period. Thus the derived policy would have the feature that many decisions (storage) in each table were equal to the unique value of the target (or the discretized value closest to that target) for that time period. The more robust the reservoir system is, the more decisive will be the target values.

The "stability" of the policies defined with release as decision variable intuitively explains the following three aspects of the simulation results in this study.

(a) As Tables 3.16 and 3.17 present, with respect to models with the previous inflow as a state variable (Model 2 and Model 4), the performance indices do not change. Model 1 performs considerably worse in Experiment 3 than in Experiment 2, as expected. However, the performance of Model 3 in Experiment 3 is unexpectedly close to the performance in Experiment 2.

This unexpected result can be explained with policy Table 3.14c. For a wide range of initial storage values (from $11\,105 \times 10^6$ m^3 to $52\,166 \times 10^6$ m^3), the release decisions are independent of the present inflow values. The errors in the present inflow forecast only affect the release decision in a very few cases (the top-right and bottom-left triangles), and the maximum deviation in decision is only one class. Therefore, the (simulated) operation based on this type of policy is insensitive to the errors in the inflow forecast.

(b) Tables 3.16 and 3.17 reveal that the models with release as a decision variable (Model 3 and Model 4) differ very little from each other with respect to all three simulated performance indices (reservoir storage, turbine release, and energy generation). In contrast, models with storage as a decision variable (Model 1 and Model 2) show a much larger difference between them with respect to performance indices.

In Table 3.14c (Model 3) and Table 3.14d (Model 4), for a wide range of initial storage values (from $12\,684 \times 10^6$ m^3 to $52\,166 \times 10^6$ m^3), the optimal decisions (release) are

Table 3.14. *Derived SDP based policy tables for the Kariba Reservoir (May)*

(a) Model 1

Initial storage	\multicolumn{6}{c}{Final storage}					
Inflow →	1	2	3	4	5	6
1	2	1	1	1	1	1
2	3	2	1	1	1	1
3	4	3	2	1	1	1
4	5	4	3	2	1	1
5	6	5	4	3	2	1
6	7	6	5	4	3	1
7	7	7	6	5	4	2
8	8	8	7	6	5	3
9	9	8	8	7	6	4
10	10	9	8	8	7	5
11	11	10	9	8	7	6
12	12	11	10	9	8	7
13	13	12	11	10	9	8
14	14	13	12	11	10	9
15	15	14	13	12	11	10
16	16	15	14	13	12	11
17	17	16	15	14	13	12
18	18	17	16	15	14	13
19	19	18	17	16	15	14
20	20	19	18	17	16	15
21	21	20	19	18	17	16
22	22	21	20	19	18	17
23	23	22	21	20	19	18
24	24	23	22	21	20	19
25	25	24	23	22	21	20
26	26	25	24	23	22	21
27	27	26	25	24	23	22
28	28	27	26	25	24	23
29	29	28	27	26	25	24
30	30	29	28	27	26	25
31	31	30	29	28	27	26
32	32	31	30	29	28	27
33	33	32	31	30	29	28
34	34	33	32	31	30	29
35	35	34	33	32	31	30
36	36	35	34	33	32	31
37	37	36	35	34	33	32
38	37	37	36	35	34	33
39	38	37	37	36	35	34
40	39	38	37	37	36	35
41	40	39	38	38	37	36
42	41	40	39	39	38	37

(b) Model 2

Initial storage	\multicolumn{8}{c}{Final storage}							
Inflow →	1	2	3	4	5	6	7	8
1	1	1	1	1	1	1	1	1
2	2	2	1	1	1	1	1	1
3	3	3	3	2	1	1	1	1
4	4	4	3	3	2	1	1	1
5	5	5	4	4	3	2	1	1
6	6	6	5	4	4	3	2	1
7	7	7	5	5	5	4	3	1
8	8	7	6	6	5	5	4	2
9	9	8	7	7	6	6	5	4
10	10	9	8	8	8	7	5	5
11	11	10	9	9	8	8	7	6
12	12	11	10	10	9	9	8	7
13	13	12	11	11	10	10	9	8
14	14	13	12	12	11	11	10	9
15	15	14	13	13	12	12	11	10
16	16	15	14	14	13	13	12	11
17	17	16	15	15	14	14	13	12
18	18	17	16	16	15	15	14	13
19	19	18	17	17	16	16	15	14
20	20	19	18	18	17	17	16	15
21	21	20	19	19	18	18	17	16
22	22	21	20	20	19	19	18	17
23	23	22	21	21	20	20	19	17
24	24	23	22	22	21	21	20	18
25	25	24	23	23	22	22	21	19
26	26	25	24	24	23	23	22	20
27	27	26	25	25	24	24	23	21
28	28	27	26	26	25	25	24	22
29	29	28	27	27	26	26	25	23
30	30	29	28	28	27	27	26	24
31	31	30	29	29	28	28	27	25
32	32	31	30	30	29	29	28	26
33	33	32	31	31	30	30	29	27
34	34	33	32	32	31	31	30	28
35	35	34	33	33	32	32	31	29
36	36	35	34	34	33	33	32	30
37	37	36	35	35	34	34	33	31
38	37	37	36	36	35	35	34	32
39	38	38	37	37	36	36	35	33
40	39	39	38	38	37	37	36	34
41	40	40	38	38	38	38	36	35
42	40	40	39	39	39	39	37	36

(c) Model 3

Initial storage	\multicolumn{6}{c}{Release}					
Inflow →	1	2	3	4	5	6
1	1	1	1	1	1	1
2	1	1	1	1	1	1
3	1	1	1	1	1	1
4	1	1	1	1	1	1
5	1	1	1	1	1	1
6	1	1	1	1	1	1
7	1	1	1	1	1	1
8	1	1	1	1	1	1
9	1	1	1	1	1	1
10	1	1	1	1	1	1
11	1	1	1	1	1	1
12	1	1	1	1	1	1
13	1	1	1	1	1	1
14	1	1	1	1	1	1
15	1	1	1	1	1	1
16	1	1	1	1	1	1
17	1	1	1	1	1	1
18	1	1	1	1	1	1
19	1	1	1	1	1	1
20	1	1	1	1	1	1
21	1	1	1	1	1	1
22	1	1	1	1	1	1
23	1	1	1	1	1	1
24	1	1	1	1	1	1
25	1	1	1	1	1	1
26	1	1	1	1	1	1
27	1	1	1	1	1	1
28	2	1	1	1	1	1
29	2	1	1	1	1	1
30	2	1	1	1	1	1
31	2	1	1	1	1	1
32	2	2	1	1	1	1
33	2	2	2	1	1	1
34	2	2	2	1	1	1
35	2	2	2	2	1	1
36	2	2	2	2	1	1
37	3	2	2	2	2	1
38	3	3	2	2	2	2
39	3	3	3	2	2	2
40	3	3	2	2	2	1
41	4	4	3	2	2	2
42	3	4	4	3	2	2

(d) Model 4

Initial storage	\multicolumn{8}{c}{Release}							
Inflow →	1	2	3	4	5	6	7	8
1	1	1	1	1	1	1	1	1
2	1	1	1	1	1	1	1	1
3	1	1	1	1	1	1	1	1
4	1	1	1	1	1	1	1	1
5	1	1	1	1	1	1	1	1
6	1	1	1	1	1	1	1	1
7	1	1	1	1	1	1	1	1
8	1	1	1	1	1	1	1	1
9	1	1	1	1	1	1	1	1
10	1	1	1	1	1	1	1	1
11	1	1	1	1	1	1	1	1
12	1	1	1	1	1	1	1	1
13	1	1	1	1	1	1	1	1
14	1	1	1	1	1	1	1	1
15	1	1	1	1	1	1	1	1
16	1	1	1	1	1	1	1	1
17	1	1	1	1	1	1	1	1
18	1	1	1	1	1	1	1	1
19	1	1	1	1	1	1	1	1
20	1	1	1	1	1	1	1	1
21	2	1	1	1	1	1	1	1
22	2	1	1	1	1	1	1	1
23	1	1	1	1	1	1	1	1
24	1	1	1	1	1	1	1	1
25	2	1	1	1	1	1	1	1
26	2	2	1	1	1	1	1	1
27	2	2	1	1	1	1	1	1
28	2	1	1	1	1	1	1	1
29	2	1	1	1	1	1	1	1
30	2	2	2	1	1	1	1	1
31	2	2	1	1	1	1	1	1
32	2	2	2	1	1	1	1	1
33	2	2	2	2	1	1	1	1
34	2	2	2	2	1	1	1	1
35	3	2	2	2	2	1	1	1
36	3	2	2	2	2	1	1	1
37	3	3	3	2	2	2	1	1
38	3	3	3	3	3	2	2	1
39	3	4	4	3	3	3	2	1
40	4	3	3	3	3	2	2	2
41	4	4	4	3	3	3	3	2
42	3	4	4	4	3	3	2	2

Table 3.15. *Simulated average annual performance (Experiment 1)*

	Model 1	Model 2	Model 3	Model 4
Indices for storage as % of reservoir capacity				
(1) Mean utilized storage	62.3%	63.7%	68.0%	67.2%
(2) Standard deviation of (1)	29.8%	21.1%	28.1%	28.4%
(3) Minimum drawdown	7.0%	24.7%	10.6%	11.0%
Indices for releases as % of annual inflow				
(4) Mean annual release	70.7%	66.5%	70.3%	70.1%
(5) Standard deviation of (4)	7.3%	8.1%	4.7%	5.2%
(6) Minimum annual release	51.3%	47.4%	57.6%	57.7%
Indices for energy as % of power capacity				
(7) Expected mean annual energy output	93.5%	86.6%	91.7%	91.4%
(8) Standard mean annual energy output	90.8%	85.5%	91.0%	90.7%
(9) Standard deviation of (8)	11.4%	11.3%	8.3%	9.0%
(10) Firm annual energy output	62.0%	59.4%	73.3%	73.8%

Table 3.16. *Simulated average annual performance (Experiment 2)*

	Model 1	Model 2	Model 3	Model 4
Indices for storage as % of reservoir capacity				
(1) Mean utilized storage	61.8%	59.8%	66.5%	66.0%
(2) Standard deviation of (1)	30.2%	21.1%	29.6%	29.2%
(3) Minimum drawdown	6.2%	22.0%	8.2%	11.1%
Indices for releases as % of annual inflow				
(4) Mean annual release	69.8%	64.5%	69.9%	69.7%
(5) Standard deviation of (4)	7.9%	7.2%	6.0%	6.2%
(6) Minimum annual release	48.4%	50.4%	55.6%	54.6%
Indices for energy as % of power capacity				
(7) Expected mean annual energy output	92.7%	87.7%	91.0%	91.1%
(8) Standard mean annual energy output	89.6%	82.4%	90.3%	90.0%
(9) Standard deviation of (8)	11.9%	10.0%	10.1%	10.2%
(10) Firm annual energy output	58.6%	62.7%	67.1%	66.5%

Table 3.17. *Simulated average annual performance (Experiment 3)*

	Model 1	Model 2	Model 3	Model 4
Indices for storage as % of reservoir capacity				
(1) Mean utilized storage	66.3%	59.8%	66.3%	66.0%
(2) Standard deviation of (1)	22.5%	21.1%	29.3%	29.2%
(3) Minimum drawdown	19.6%	22.0%	10.4%	11.2%
Indices for releases as % of annual inflow				
(4) Mean annual release	67.2%	64.5%	69.9%	69.7%
(5) Standard deviation of (4)	5.1%	7.2%	6.3%	6.2%
(6) Minimum annual release	58.8%	50.4%	54.7%	54.6%
Indices for energy as % of power capacity				
(7) Expected mean annual energy output	92.7%	87.7%	91.0%	91.1%
(8) Standard mean annual energy output	86.4%	82.4%	90.3%	90.0%
(9) Standard deviation of (8)	7.5%	10.0%	10.2%	10.2%
(10) Firm annual energy output	73.3%	62.7%	66.66%	66.5%

independent of both the present and the previous inflow values. The differences in the present or the previous inflows only affect the release decisions in a very few cases (the top-right and bottom-left triangles). The maximum deviation in decision is one class. In contrast, Table 3.14a (Model 1) and Table 3.14b (Model 2) show that the optimal decisions (final storage) for the operation are strongly determined by the initial storage values and the inflow values. A small variation in the inflow value can lead to a different decision. Therefore, whether the previous or the present inflow is used as a state variable strongly influences the decisions during operations (and hence, the performance indices). Comparing the structures of Table 3.14c and Table 3.14d with those of Table 3.14a and Table 3.14b, it can be concluded that the policies derived from models with release as a decision variable (Model 3 and Model 4) are much less sensitive to variations in the initial storage and inflow than policies derived from models with storage as a decision variable (Model 1 and Model 2).

(c) Tables 3.15 to 3.17 (particularly Table 3.17, when the imperfect inflow forecast is adopted as a guide during operation simulation) indicate that the reservoir performances obtained from the models with release as a decision variable (Model 3 and Model 4) are better than those obtained from the models with storage as a decision variable (Model 1 and Model 2). Model 3 and Model 4 result in larger mean values and smaller fluctuations in energy generation, in both annual and monthly performance indices. These results can be related to the "stability" of policy Table 3.14c and Table 3.14d.

From this discussion, it can be concluded that the models with release as the decision variable considerably outperform the models with storage as the decision variable. If the "right" decision has been made regarding the decision variables, the different choices of inflow state variables would not much affect the performance of the system. However, in reality the selection of the "right" decision variable cannot always be realized. For example, for multipurpose reservoir systems, sometimes more than one objective has to be optimized at the same time. Some objectives might be directly related to release and others might be directly related to storage. As has been shown in the case study, when storage is selected as a decision variable, models become sensitive to the choice of inflow state variable.

In Experiment 3, a simple linear regression model (Budhakooncharoen, 1986) with considerable errors (see Table 3.12) forecasts the inflow. Even so, the Model 1 based simulation still performs better than the Model 2 based simulation that depends on the previous inflow. This makes the model with present inflow as a state better than the

model with previous inflow as a state. Although it cannot be concluded from the current results that this type of model with present inflow as a state will always outperform the type of model with previous inflow as a state, it does illustrate that the policy derived from Model 1 can sustain errors in the inflow forecast. Besides, in real-time operation, having large amounts of up-to-date information regarding rainfall, river channel flow, ground water and catchment area characteristics, etc., a good inflow forecast can be easily produced.

3.3.3 Inflow serial correlation assumptions

Serial correlation or autocorrelation means that the value of the stochastic variable under consideration at one time period is correlated with the values of the stochastic variable at earlier periods. The correlation between an observation at a certain time period with an observation k time periods earlier is called the kth order serial correlation.

The serial assumption for the stochastic inflow sequences is an important issue in reservoir operation optimization. In SDP models, the serial correlation assumption is used to describe the inflow sequence. The stochastic nature of the inflow sequence is a generally observed fact. However, the choice of the serial correlation assumption is an unsolved controversy in the literature of reservoir operation optimization.

When the SDP model was first introduced into reservoir operation, the inflow sequence was assumed as a Markov-I process. Later, the independence assumption was also used in SDP models. The Markov-I assumption is more popular than the independence assumption. But supporters of the use of each of these assumptions have presented arguments to favor one above the other. These arguments are often supported by experimental results on reservoir operation optimization either for different real problems or with different SDP model setups (e.g., different discretization, different decision or state variables, different objectives, etc.). Thus the results are not often comparable. The discussion on which inflow assumption is the best remains undecided.

This section presents the influence that different inflow serial correlation assumptions have on the performance of SDP models. To obtain an overview of the problem, besides models with the Markov-I and the independence assumptions, another two models are presented. One model considers the serial correlation one step further than the Markov-I assumption: the SDP model with Markov-II assumption. The other model interprets the inflow process even more simply than the independence assumption does: the model with the assumption that the inflow is deterministic.

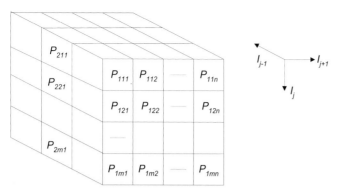

Figure 3.6 Graphical illustration of the three-dimensional (Markov-II) transition probabilities

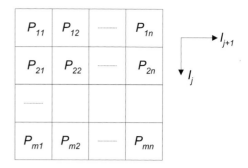

Figure 3.7 Graphical illustration of the two-dimensional (Markov-I) transition probabilities

THE FOUR SERIAL CORRELATION ASSUMPTIONS AND THEIR MODELING COMPLEXITIES

Many river flow time series exhibit serial correlation. That is, high flows follow high flows and low flows follow low flows. This phenomenon is particularly evident for short time intervals. Annual and seasonal flows (the total flow amount of an entire period) are seldom highly correlated, while monthly, weekly, and especially daily or hourly flows generally exhibit high serial correlations.

For the application of the SDP model in the optimization of reservoir operation, the inflow serial correlation is interpreted by transition probabilities. These transition probabilities are coupled with the recursive relation of DP to derive expectation-oriented optimal values. The formulations for the SDP models with Markov-II, Markov-I, independence, and deterministic inflow serial correlation assumptions are given in this section. The modeling and computational complexities of these models are analyzed. In all four models, the present inflow is selected as the inflow state variable.

MARKOV-II INFLOW PROCESS ASSUMPTION

The Markov-II assumption takes lag-two and lag-one serial correlations of the inflow process into consideration. The transition probability of the Markov-II process can be characterized as

$$P_{j+1}(I_{j+1}|I_j, I_{j-1}, I_{j-2}, \ldots) = P_{j+1}(I_{j+1}|I_j, I_{j-1}). \quad (3.26)$$

The SDP model with Markov-II assumption can be formulated with the following recursive relation:

$$F_j^n(k, m, p) = \operatorname*{Opt}_l\left[B_{k,p,l,j} + \sum_q P_{m,p,q}^{j+1} \times F_{j+1}^{n-1}(l, p, q)\right]. \quad (3.27)$$

Time notations and variables presented in Section 3.3.2 are used.

The transition probabilities, $P_{m,p,q}^{j+1} = P_{j+1}(I_{j+1}|I_j, I_{j-1})$, of a Markov-II process can be represented as a three-dimensional array. Figure 3.6 shows a graphical illustration of such a three-dimensional array.

MARKOV-I INFLOW PROCESS ASSUMPTION

The Markov-I assumption takes the lag-one (first order) serial correlation of the inflow process into consideration. The transition probability of the Markov-I process can be expressed as

$$P_{j+1}(I_{j+1}|I_j, I_{j-1}, I_{j-2}, \ldots) = P_{j+1}(I_{j+1}|I_j). \quad (3.28)$$

The SDP model with Markov-I assumption can be formulated with the following recursive relation:

$$F_j^n(k, p) = \operatorname*{Opt}_l\left[B_{k,p,l,j} + \sum_q P_{p,q}^{j+1} \times F_{j+1}^{n-1}(l, q)\right]. \quad (3.29)$$

The transition probabilities, $P_{p,q}^{j+1} = P_{j+1}(I_{j+1}|I_j)$, of a Markov-I assumption can be represented as a two-dimensional array as shown in Figure 3.7. It can be considered as a special case of the transition probabilities of a Markov-II process that each layer along the axis of I_{j-2} has a unique probability distribution.

INDEPENDENCE OR RANDOM INFLOW PROCESS ASSUMPTION

The inflow process is considered as exhibiting no serial correlation with inflows of the previous time periods with the independence or random inflow process assumption. That is, the probability is independent of the previous inflow station:

$$P_{j+1}(I_{j+1}|I_j, I_{j-1}, I_{j-2}, \ldots) = P_{j+1}(I_{j+1}). \quad (3.30)$$

Figure 3.8 Graphical illustration of the one-dimensional (independence) transition probabilities

The SDP model with the independence inflow assumption can be formulated with the following relation:

$$F_j^n(k,p) = \underset{l}{\text{Opt}} \left[B_{k,p,l,j} + \sum_q P_q^{j+1} \times F_{j+1}^{n-1}(l,q) \right]. \quad (3.31)$$

The transition probabilities resulting from the independence inflow assumption can be represented as a one-dimensional array as shown in Figure 3.8. It can be considered as a special case of the transition probabilities of the Markov-I assumption that each line along the axis of I_j has a unique probability distribution.

DETERMINISTIC INFLOW PROCESS ASSUMPTION

The deterministic assumption assumes that there is a predetermined inflow I_j, for each time period j, thus:

$$P_{j+1}(I_{j+1}|I_j, I_{j-1}, I_{j-2}, \ldots) = P_j(\overline{I_j}) = 1.0. \quad (3.32)$$

The mean inflow value of a time period j (e.g., the average value of inflows of the month over a number of years) is used as I_j.

The (deterministic) model with the deterministic inflow assumption can be expressed in the following recursive relation:

$$F_j^n(k) = \underset{l}{\text{Opt}} \left[B_{k,p,l,j} + F_{j+1}^{n-1}(l) \right], \quad (3.33)$$

where $F_j^n(k)$ is the (sub)optimal value of the recursive equation at stage n (period j) as a function of S_j.

The deterministic assumption may be considered as a special case of the independence inflow process assumption. In that case the inflow of each time period is discretized into only one class and thus the occurrence probability is 1.0.

Next, the complexities involved in the modeling of these four types of models are discussed and compared with each other.

First, consider the Markov-II assumption. From Figure 3.6 it can be seen that if the inflow is divided into m classes for each time period, then the total number of transition probabilities is equal to m^3. These m^3 transition probabilities of the inflow process have to be estimated from historical inflow data. To estimate such a large number of transition

probabilities (parameters), not only many calculations have to be performed, but more importantly a large estimation error may occur on the transition probabilities. The transition probability, $P_{j+1}(I_{j+1} | I_j, I_{j-1})$, is calculated as given below.

$P_{j+1}(I_{j+1} | I_j, I_{j-1}) =$ occurrence frequency of I_{j+1}, at time period $j+1$ given the inflows are I_j and I_{j-1} at time period j and $j-1$, respectively.

If the total number of available historical observation data is not very large, say fewer than $1/2 \times m^3$, then most of the transition probabilities will be zero as there are not enough occurrences. Thus, it requires a large number of historical observation data to obtain a reasonably accurate estimation of the transition probabilities. This is however a big difficulty, because historical inflow data covering more than 30 or 40 years are scarcely available. Also, consider the computational complexity of an SDP model with the Markov-II assumption. Equation (3.26) indicates that the number of evaluations is proportional to m^3.

Second, consider the Markov-I assumption. From Figure 3.7 it is clear that if the inflow is divided into m classes, then the total number of transition probabilities is equal to m^2. Thus, the number of transition probabilities to be estimated is a factor of m less than that with the Markov-II assumption. Consequently, with the same amount of historical inflow data, the estimation of the transition probabilities will be more accurate than the estimation of the Markov-II transition probabilities. Furthermore, from Eq. (3.28) it can be seen that the number of evaluations is proportional to m^2. Thus, the computational complexity of a SDP model with the Markov-I assumption is a factor m less than that of a SDP model with the Markov-II assumption.

Similar analysis can be made for the SDP model with the independence inflow assumption and the (deterministic) model with the deterministic inflow assumption. The number of transition probabilities to be estimated is m for the independence assumption (Eq. 3.30) and is 1 for the deterministic inflow assumption (Eq. 3.32).

BEST SERIAL CORRELATION ASSUMPTION

The Markov-I assumption has been adopted by many researchers to model the inflow. The independence inflow process assumption has also been used by some researchers (Su and Deininger, 1974; Laabs and Harboe, 1988; Huang et al., 1991). However, the simpler model with the independence inflow process assumption has never enjoyed the same popularity that the Markov-I assumption has in the application of SDP in reservoir operation. The independence assumption was criticized as too simple to describe the stochasticity of inflow time series accurately. However, this argument is not sufficient to decide which inflow assumption should be used in the SDP models. In fact, it is important to have an inflow assumption that reflects the real nature of inflows. It is also very important not to make the model unnecessarily complicated. The best model should be the simplest model which still "sufficiently" reflects the reality. The term "sufficient" depends on the application. For SDP models applied to reservoir operation optimization, a model is sufficiently good if it produces approximately the same performance as the best (complicated) models.

The modeling and computational complexity of SDP models decrease by a factor m from Markov-II, Markov-I, independence inflow to deterministic inflow assumption as shown previously.

With respect to the errors in the SDP model caused by the inflow assumption, two types of errors can be distinguished. One is when the model is too simple to describe those properties of the natural phenomena that are important for the decision. The other is when the set of historical samples is too small. The error caused by a small sample set increases as the complexity of the model increases (the number of parameters to be estimated increases). The difficulty in making a good choice among the serial correlation assumptions is mainly due to the difficulty in determining the appropriate interchange between these two types of errors.

The question about which serial correlation assumption gives the best SDP model cannot be easily answered. It depends on many factors, and thus there may be no single best SDP model. It depends on the situation to which the SDP models are applied. Therefore, this problem is further analyzed through several case studies.

The SDP models have been mainly applied in reservoir operations for their long-term management. The time period usually is a month or half a month or so. For this level of time it is not uncommon to have systems with low serial correlation for many time periods within a year. To illustrate this, the monthly serial correlation coefficients of the three available inflow time series from the systems Kariba and Victoria/Randenigala are listed in Table 3.18.

From the serial correlation coefficients in Table 3.18, the following facts can be observed. For many months, the inflow series of the Kariba system is highly serially correlated (both lag-one and lag-two). The inflow series of the Mahaweli (Victoria/Randenigala) system can be considered as moderately serially correlated.

INVESTIGATION OF SUITABILITY OF CORRELATION ASSUMPTIONS

This section presents five experiments carried out to investigate the pros and cons of the four serial correlation assumptions based on the Kariba Reservoir and the Victoria/Randenigala Reservoirs. Table 3.19 gives a summary of the

Table 3.18. *Serial correlation coefficients of the three case study systems*

Month	Kariba		Victoria	Randenigala
	lag-one	lag-two	lag-one	lag-one
1	0.36	0.07	0.22	0.22
2	0.39	0.22	0.29	0.07
3	0.10	0.42	0.50	0.36
4	0.52	0.11	0.39	0.21
5	0.65	0.45	0.51	0.53
6	0.53	0.55	0.68	0.55
7	0.89	0.31	0.31	0.44
8	0.92	0.82	0.55	0.49
9	0.93	0.79	0.19	0.36
10	0.97	0.97	0.07	0.11
11	0.94	0.88	0.01	0.05
12	0.79	0.77	0.54	0.33

five experiments. Detailed design and the results of each experiment are presented subsequently.

EXPERIMENT A: KARIBA SYSTEM OPERATION BASED ON MODELS WITH RELEASE AS DECISION VARIABLE

In this experiment the Kariba Reservoir, which has relatively high inflow serial correlation coefficients (see Table 3.18) is selected as the case study.

The experiment aims to screen the performance of the SDP models with the four different inflow assumptions: Markov-II (Eq. 3.27), Markov-I (Eq. 3.29), independence (Eq. 3.31), and deterministic (Eq. 3.33). Based on the conclusion of Section 3.3.2, the release (which is more directly related to the objective of energy generation) is selected as the decision variable. The decision variable in Eqs. (3.27), (3.29), (3.31), and (3.33) can now be specified as R_j.

In this experiment the structure of the SDP model is similar to the model described by Eq. (3.24), except for different inflow assumptions. The stage is the time step, which is one month. The objective is to maximize the expected annual energy generation. The optimization is subjected to the physical constraints of the reservoir system (e.g., storage constraints, release constraints, etc.). The reservoir storage is discretized into 42 classes with equal size. Release levels up to twice the monthly release capacities of the penstocks are to be optimized. They are discretized into 6 classes with equal size. For the three stochastic inflow assumptions the monthly inflows are discretized into varying numbers of classes (from 2 to 8, according to the discretization range of the 24 years historical monthly inflow; 1961–84) with equal occupancy

Table 3.19. *Key points of the design of experiments*

	Experiment A	Experiment B	Experiment C	Experiment D	Experiment E
Case study	Kariba	Kariba	Kariba	Kariba	Victoria/Randenigala
Inflow assumption of model	(a) Markov-II (b) Markov-I (c) Independence (d) Deterministic	(a) Markov-II (b) Markov-I (c) Independence (d) Deterministic	Markov-I if (i) 0.0 (ii) 0.5 (iii) 0.75 (iv) 0.9 (v) 1.0 Otherwise independence	(a) Markov-I (b) Independence	(a) Markov-I (b) Independence
Decision variable	Release	Final storage	Final storage	Final storage	Final storage
State variable	Present inflow, initial storage	Present inflow, initial storage	Present inflow, initial storage	Present inflow, initial storage	Present inflow, initial storage
Objective	Max. expected annual energy	Max. expected annual energy	Max. expected annual energy	Max. expected annual energy	Max. expected annual energy
Constraint	—	—	—	—	Irrigation demand
Simulation	According to policy based on perfect forecast	According to policy based on perfect forecast	According to policy based on perfect forecast	According to policy based on imperfect forecast	According to policy based on perfect forecast

Table 3.20. *Simulated average annual performance (Experiment A)*

	Markov-II	Markov-I	Independence	Deterministic
Indices referring to energy as % of power capacity				
(1) Expected mean annual energy output	91.3%	91.7%	96.6%	95.6%
(2) Simulated average annual energy output	91.0%	91.0%	90.6%	90.0%
(3) Standard deviation of (2)	8.3%	8.3%	10.0%	9.4%
(4) 95% confidence interval for mean energy	(86.5%, 95.5%)	(86.5%, 95.5%)	(85.2%, 96.0%)	(84.9%, 95.1%)
(5) Minimum annual energy output	73.7%	73.3%	69.6%	70.0%
Indices referring to storage as % of reservoir capacity				
(6) Average utilized storage	68.1%	68.0%	65.2%	67.1%
(7) Standard deviation of (6)	27.7%	28.1%	29.4%	30.0%
(8) Minimum drawdown	11.9%	10.6%	8.6%	10.8%
Indices referring to release as % of annual inflow				
(9) Average annual release	70.3%	70.3%	70.3%	69.6%
(10) Standard deviation of (9)	4.7%	4.7%	6.0%	5.5%
(11) Minimum annual release	60.7%	57.6%	57.6%	57.7%

frequencies. For the deterministic inflow assumption, the monthly inflows are "discretized" into one class. The median of each inflow class is the representative value of that class. With the derived SDP based optimal operation policies, the performances of the reservoir system were simulated with the 12 years (1973–84) historical inflow time series. Perfect forecasting is assumed to be available at the beginning of each time period. The simulations "strictly" rely on the derived optimal operation policies, as long as the physical constraints of the reservoir system are not violated. Table 3.20 presents the simulated average annual performance indices from Experiment A.

Table 3.21. *Simulated average annual performance (Experiment B)*

	Markov-II	Markov-I	Independence	Deterministic
Indices referring to energy as % of power capacity				
(1) Expected mean annual energy output	93.0%	93.7%	99.9%	99.2%
(2) Simulated average annual energy output	90.9%	90.8%	90.9%	80.1%
(3) Standard deviation of (2)	11.3%	11.4%	12.9%	17.8%
(4) 95% confidence interval for mean energy	(84.8%, 97.0%)	(84.6%, 97.0%)	(83.9%, 97.9%)	(70.5%, 89.7%)
(5) Minimum annual energy output	62.2%	62.0%	58.4%	48.7%
Indices referring to storage as % of reservoir capacity				
(6) Average utilized storage	62.4%	63.3%	62.0%	52.9%
(7) Standard deviation of (6)	29.5%	29.8%	30.9%	6.0%
(8) Minimum drawdown	8.5%	7.0%	3.1%	43.7%
Indices referring to release as % of annual inflow				
(9) Average annual release	70.7%	70.7%	70.7%	63.4%
(10) Standard deviation of (9)	7.3%	7.3%	8.5%	14.5%
(11) Minimum annual release	51.2%	51.3%	48.6%	38.2%

Simulated performance is presented according to the following three aspects: (a) energy generation, (b) reservoir storage, and (c) releases through the turbine. For each of the three performance indices, the simulated mean, the standard deviation, and the minimum value are presented. For energy generation, the expected annual gain obtained from the SDP based optimization is also presented. The "expected annual energy output" itself does not tell much about the real performance of the system. However, the difference between the expected value (item 1 in Table 3.20) and the simulated value (item 2) shows how far the optimization model is from the real nature of the problem being optimized.

According to the results, although the ways of inflow being considered differ very much for the four optimization models, the simulated performances based on the four derived "optimal" policies differ very little. For example, for the objective value of annual energy output (item 2), the smallest simulated energy output from the deterministic inflow assumption is only 1% less than the largest output from Markov-II and Markov-I assumptions.

The present result can be understood by recalling the conclusion of Section 3.3.2 regarding the influence of the decision variable of SDP. When the variable directly related to the objective of the optimization is selected as the decision variable, the model becomes insensitive to the way in which the inflow is considered. In this situation, simpler models (either the SDP model with independence inflow or even the deterministic model based on mean value of inflow) would perform almost as well as the more complicated models that consider inflow serial correlations.

EXPERIMENT B: KARIBA SYSTEM OPERATION BASED ON MODELS WITH STORAGE AS DECISION VARIABLE

This experiment also aims to screen the performance of the SDP models with the four different inflow assumptions: Markov-II (Eq. 3.27), Markov-I (Eq. 3.29), independence (Eq. 3.31), and deterministic (Eq. 3.33). As in Experiment A, the Kariba Reservoir is selected as the case study in this experiment. The final storage is chosen to be the decision variable in the present experiment. Therefore, the decision variable in Eqs. (3.27), (3.29), (3.31), and (3.33) can now be specified as S_{j+1}. The balance setup of the SDP models and the path of subsequent simulations remain the same as in Experiment A.

Table 3.21 presents the simulated average annual performance indices from Experiment B. Similar to Table 3.20, they are presented according to the following three aspects: (a) energy generation, (b) reservoir storage, and (c) releases through the turbine.

From the results of the present experiment, the influence of the SDP model with different inflow assumptions can be much better detected than from Experiment A.

First, the policy derived from the model with the deterministic inflow assumption leads to a considerably worse performance of the reservoir system as compared with that from the models with the stochastic inflow assumption. For example, the simulated mean annual energy output (item 2) resulting from the model with the deterministic inflow assumption is about 12% lower than that of the models with stochastic inflow assumptions. The fluctuation or standard deviation of annual energy generation (item 3) is almost 50% higher than that for the other three models. The firm annual energy output (item 4) is about one-quarter less than that of the other three models.

This result clearly illustrates the drawback of the model with the deterministic inflow assumption in deriving reservoir operation policies. The deterministic model based on the mean value of inflows seems too simple to represent the nature of reservoir inflows sufficiently. In general, a deterministic model is probably a good tool to screen the best performance a system could have if the historical inflow observations (which are already known) repeat in the future. If the issue is to derive reservoir operation policies, the deterministic model probably may function better within the framework of the so-called "implicit" type stochastic approach.

Second, the policy derived from the SDP model with the Markov-II assumption does not show much improvement in the reservoir system performance as compared with that of the model with the Markov-I assumption. Compare the two columns of performance indices corresponding to the Markov-II and Markov-I assumptions. It seems that except for the unimportant indices referring to the reservoir minimum drawdown (item 8), the differences of all the indices are less than 1%.

This result indicates that the improvement resulting from the model with the Markov-II assumption does not justify the additional complication of the model. Generally, SDP models with second or higher order inflow serial correlation assumptions are not practical. The difficulty lies in the estimation of the three-dimensional inflow transitional probabilities. The availability of observed data of inflow time series for over 40 years is scarce. Such a limited length of historical inflow time series is bound to result in considerable errors in the estimation of three-dimensional inflow transition probabilities. Those errors may significantly diminish the merit of the Markov-II based model even when it better reflects the characteristics of the inflow.

When comparing the performance between the Markov-I and independence inflow assumptions, a firm conclusion is somewhat difficult to draw based on the present results. The objective values of mean annual energy outputs (item 2) resulting from both assumptions are almost similar. The model with the Markov-I assumption leads the system to slightly better performance in the sense of the standard deviation of annual energy output (item 3) and firm annual energy output (item 4). However, as compared with the model with the deterministic inflow assumption, the model with the independence assumption does not vary considerably from the model with the Markov-I assumption.

EXPERIMENT C: KARIBA SYSTEM OPERATION BASED ON MODELS WITH MARKOV-I AND INDEPENDENCE ASSUMPTIONS

The experiment aims to obtain additional insight into the performance of the SDP models with the Markov-I and independence assumptions based on the Kariba system.

Table 3.18 shows that the correlation coefficients of the Kariba inflows are high for some months and low for the others. The idea of using a model with the Markov-I assumption for the months with high correlation coefficients and the independence assumption for the months with low correlation coefficients may be a reasonable choice for such a system. This type of model can reflect the serial correlation of the inflow time series for the months when the serial correlation is high. It also can avoid the unnecessary additional parameter estimation errors for the months when the serial correlation is low.

To investigate this idea, five models are set up in the present experiment. For each model a critical point (cp) is defined. They are (i) 0.0, (ii) 0.5, (iii) 0.75, (iv) 0.9, and (v) 1.0. For the months with lag-one serial correlation coefficients larger than or equal to the critical point, the Markov-I transition probabilities will be coupled into the recursive relation of the SDP model (Eq. 3.29). For the months with lag-one serial correlation coefficients smaller than the critical point, the independence probabilities will be coupled into the recursive relation of the SDP model (Eq. 3.31). Consider Model (iii) (Table 3.19) as an example. There are 6 months (from month 7 to month 12) whose lag-one correlation coefficients are larger than 0.75 (see Table 3.18). Therefore, for those 6 months the Markov-I transition probabilities and for the remaining 6 months the independence probabilities will be coupled into the recursive relation. Similarly, for Model (ii) there will be 9 months (from month 4 to 12) with Markov-I transition probabilities and 3 months with independence probabilities. For Model (iv) there will be 4 months (from month 8 to 11) with Markov-I transition probabilities and 8 months with independence probabilities. Model (i) with 0.0 as critical point is the model with the Markov-I assumption for all the 12 months in a year. Model (v) with 1.0 as critical point is the model with the independence assumption for all the 12 months in a year. The balance setup of the SDP models and the procedure adopted in the subsequent simulations are the same as in Experiment B. Table 3.22 presents the simulated annual performance indices from Experiment C.

Table 3.22 shows that the variations among the simulated performances resulting from the five models are very limited. For the indices for storage, gradual minor changes can be observed from the model with the Markov-I inflow assumption (cp = 0.0) to the model with the independence assumption (cp = 1.0), with the increase of critical point value. For the

Table 3.22. *Simulated average annual performance (Experiment C)*

	cp = 0.0 Markov-I	cp = 0.50	cp = 0.75	cp = 0.90	cp = 1.0 Independence
Indices referring to energy as % of power capacity					
(1) Expected mean annual energy output	93.7%	93.7%	95.3%	97.3%	99.9%
(2) Simulated average annual energy output	90.8%	90.8%	91.6%	91.0%	90.9%
(3) Standard deviation of (2)	11.4%	11.44%	12.3%	12.1%	12.9%
(4) 95% confidence interval for mean energy	(84.6%, 97.0%)	(84.6%, 97.0%)	(84.3%, 97.7%)	(84.4%, 97.6%)	(83.9%, 97.9%)
(5) Minimum annual energy output	62.0%	62.0%	58.6%	61.9%	58.4%
Indices referring to storage as % of reservoir capacity					
(6) Average utilized storage	63.3%	63.3%	62.3%	62.2%	62.0%
(7) Standard deviation of (6)	29.8%	29.8%	30.2%	30.5%	30.9%
(8) Minimum drawdown	7.0%	7.0%	5.8%	4.5%	3.1%
Indices referring to release as % of annual inflow					
(9) Average annual release	70.7%	70.7%	70.8%	70.8%	70.7%
(10) Standard deviation of (9)	7.3%	7.3%	8.1%	7.9%	8.5%
(11) Minimum annual release	51.3%	51.3%	48.4%	51.4%	48.6%

indices for energy and release, a very small jump at the critical point 0.9 can be detected. The standard deviations (items 3 and 10) are smaller and the minimum values (items 5 and 11) are bigger at the critical point 0.9 when compared with the two neighboring points, 0.75 and 1.0. This can be interpreted as a positive sign for the idea of using a model with the Markov-I assumption for the months with high correlation coefficients and the independence assumption for the months with low correlation coefficients. However, the improvement is too small to justify the additional complications involved with the model.

EXPERIMENT D: KARIBA SYSTEM OPERATION BASED ON IMPERFECT FORECAST

From the results of both Experiments B and C, it seems that the model with the Markov-I assumption leads the Kariba system to a slightly better performance when perfect forecasting is available (as used in the experiments). However, the policy derived from the model with the Markov-I inflow assumption is likely to be more sensitive to the accuracy of inflow forecasting. In real-time operation when the inflow forecasting is not perfect, the trade-off between the two SDP models with the Markov-I and independence inflow assumptions may be different.

The present experiment aims to obtain insight into the performance of SDP models with the Markov-I and independence inflow assumptions when the inflow forecasting is not perfect during operation simulation. The reservoir system in consideration is still the Kariba system.

The setup of the SDP models is the same as the two models (Markov-I and independence) in Experiment C. The path of subsequent simulations differs from Experiment C. The derived optimal policies are implemented at the beginning of each time period with forecast inflow instead of the actual inflow. The inflows are forecast according to the readily available regression analysis of Budhakooncharoen (1986). Table 3.23 presents the simulated annual performance indices from Experiment D.

Table 3.23 shows that the differences between all the simulated performance indices resulting from the two models (with Markov-I and independence assumptions) become even smaller, as compared with those when perfect inflow forecasting is available (see Table 3.22, the models with Markov-I and independence assumptions). As for the simulated mean annual energy output (item 2), the model with the independence assumption slightly outperforms the model with the Markov-I assumption. As for the firm annual energy output (item 4), the model with the Markov-I assumption slightly outperforms the model with the independence assumption. The standard deviations of mean annual energy output (item 3) are the same for the two models. The impression obtained from Table 3.23 is that, when the inflow forecasting is not perfect, it is difficult to identify the most suitable model (with Markov-I or independence assumptions) for the Kariba system with respect to energy production.

From the results of Experiments B, C, and D, it can be observed that the model with the Markov-I assumption

Table 3.23. *Simulated average annual performance (Experiment D)*

	Markov-I	Independence
Indices referring to energy as % of power capacity		
(1) Expected mean annual energy output	93.7%	99.9%
(2) Simulated average annual energy output	86.6%	87.1%
(3) Standard deviation of (2)	7.4%	7.4%
(4) 95% confidence interval for mean energy	(82.6%, 90.6%)	(83.1%, 91.1%)
(5) Minimum annual energy output	76.5%	73.6%
Indices referring to storage as % of reservoir capacity		
(6) Average utilized storage	67.1%	66.2%
(7) Standard deviation of (6)	24.2%	24.7%
(8) Minimum drawdown	14.8%	14.3%
Indices referring to release as % of annual inflow		
(9) Average annual release	67.2%	67.7%
(10) Standard deviation of (9)	4.6%	4.2%
(11) Minimum annual release	60.4%	61.0%

leads the Kariba system to a slightly better performance when perfect forecasting is available (as defined in Experiments B and C). However, when considering the additional complexity of the SDP model with the Markov-I assumption, the improvement does not seem to be substantial. Besides, in real-time operation, when the inflow forecasting is not perfect, the small improvement of the model with the Markov-I assumption will diminish. From these points of view, the SDP model with the independence inflow assumption can be termed a more suitable one than the SDP model with the Markov-I assumption for a system like Kariba with high inflow serial correlation coefficients for many months but with short observed inflow data.

EXPERIMENT E: MAHAWELI SYSTEM OPERATION

This experiment aims to compare the model with the Markov-I assumption with the model with the independence inflow assumption. In the experiment, the two-unit reservoir system of Victoria/Randenigala, whose inflow time series has a low serial correlation compared with Kariba, is selected as the case study.

The setup of the SDP models is similar to that described in Section 3.3.1 (Table 3.6), except for the different inflow assumptions. The recursive relations of the two models are defined as in Eq. (3.29) and Eq. (3.31), respectively, while the decision variable in the equations is final storage S_{j+1}. The stage is the time step, which is one month. The objective is to maximize the expected annual energy generation subject to the constraints of satisfying downstream irrigation requirements. The 32 years (1949–80) of available observed inflow

data record is used to obtain statistical parameters of the stochastic inflow. Four inflow classes and seven storage classes with equal size intervals have been considered for both reservoirs in cascade, thus yielding $4 \times 4 = 16$ inflow class combinations and $7 \times 7 = 49$ storage class combinations. The median of each inflow class is the representative value of that inflow class. In the subsequent simulations the historical inflow time series have been used "strictly" relying on the derived optimal operation policies, as long as the physical constraints of the reservoir system are not violated. During the simulation, perfect forecasting is assumed to be available at the beginning of each time period. Table 3.24 presents the simulated performance indices from Experiment E.

The simulated performance is presented in two aspects: (a) energy generation, and (b) irrigation supply. For energy generation, the simulated mean, the standard deviation, and the minimum value are presented. The optimization of this system does not hold for the whole set of decisions in the annual cycle due to the constraint of irrigation demands (see Section 3.3.1). Therefore, the expected annual energy output is not obtained. The table presents the performance indices of reliability (both time-based and quantity-based), repairability, and vulnerability for the irrigation supply.

The performance indices of energy output in Table 3.24 indicate that the model with the Markov-I inflow assumption leads to slightly better system performance (e.g., larger minimum annual energy, item 4) compared with the independence assumption. However, the indices of irrigation supply imply that the model with the independence inflow assumption leads to slightly better system performance (e.g., smaller reparability

Table 3.24. *Simulated performance (Experiment E)*

	Markov-I	Independence
Indices referring to energy as % of power capacity		
(1) Simulated average annual energy output	52.5%	52.2%
(2) Standard deviation of (1)	11.6%	11.6%
(3) 95% confidence interval for mean energy	(49.0%, 56.0%)	(48.7%, 55.7%)
(4) Minimum annual energy output	31.7%	28.2%
Indices referring to irrigation supply		
(5) Time-based reliability[a]	86.2%	86.2%
(6) Quantity-based reliability[b]	95.9%	96.0%
(7) Repairability (month)[c]	1.57	1.47
(8) Vulnerability (10^6 m^3)[d]	60.5	56.1

[a] % of time-based steps with fulfilled irrigation demand
[b] % of accumulated irrigation demand met
[c] Average duration of an irrigation failure (shortage) event
[d] Average accumulated irrigation shortage per failure

in item 7 and vulnerability in item 8) compared with the other. Where the most important indices (mean annual energy and reliability of irrigation supply) are concerned, there is hardly any difference between the two models.

The results from the experiment show that the two SDP models (with Markov-I and independence assumptions) lead the Victoria/Randenigala system to almost equal utilization of water in the reservoirs when perfect inflow forecasting is available. Therefore, the SDP model with the independence inflow assumption can be considered to be better than the SDP model with the Markov-I assumption due to its simplicity in modeling.

3.3.4 Summary of observations

The analysis of the characteristics of a Markov chain and the convergence behavior of the SDP model show that the large number of zero elements in transition probability matrices seems to be the cause of failing to satisfy the convergence criterion, stabilization of expected annual increment of the objective function value, in the SDP model. The study shows

that the substitution of these zeros with reasonably small values is a suitable method to overcome the above problem.

The study carried out to investigate the influence of different decision variables and inflow state variables on the performance of the SDP model, based on several versions of the SDP model, shows that the variable directly related to the objective of optimization is to be preferred as the decision variable. The choice of the inflow state variable considerably affects the operation of the system if the selected decision variable is not directly related to the objective of optimization.

The suitability of different inflow serial correlation assumptions in the SDP model has been examined through models formulated based on Markov-I, Markov-II, independence, and deterministic inflow assumptions. The analysis indicated that the SDP model becomes insensitive to the above inflow assumptions if the selected decision variable is directly related to the objective of optimization. A comparison among the above assumptions has been made based on the complexity involved in the computations, the length of inflow time series available, time step length considered in optimizations, and errors possible in inflow forecast.

4 Optimal reservoir operation for water quality

When a flowing river is dammed and becomes an impoundment, two major changes occur. First, creating an impoundment greatly increases the time required for water to travel the distance from the headwaters to the discharge at the dam. Second, thermal or density and therefore chemical stratification may take place. Both have a marked effect on water quality. Both the increased detention time and thermal stratification in an impoundment change the characteristics of the water discharged at a given geographical location from what they were when the stream was free flowing. Some effects of impoundments improve water quality; others deteriorate it. This also implies the possibility of using the reservoirs for control of the quality of water besides merely satisfying the quantity requirement.

The increased emphasis on water quality accents the need for formulation of methodologies for operating reservoirs for control of water quality. Considering reservoir dynamics while applying optimization techniques for operational decisions enables policies for a reservoir accounting for the quality of water supplied besides satisfying quantity requirements. The assumption of complete instantaneous mixing of water in a reservoir throughout its entire volume is an over-simplification compared to the real behavior of reservoirs that undergo mixing and stratification cycles. This chapter presents models that assume complete mixing in reservoirs while deriving optimum operation policies when quality aspects are of interest.

There have been relatively few studies of optimum reservoir operation in which water quality has been considered. However, due to the increasing demand for water of good quality, consideration of the quality aspects in reservoir operation optimization has become very important.

Verhaeghe and Tholan (1983) analyzed an optimal water allocation problem satisfying both quantity and quality objectives. The objective was to minimize the economic losses that occur due to water shortages and poor water quality. Salinity characterized the water quality. The same problem, allocation of water from a river to three irrigation areas via reservoirs, was formulated into four problems having different schematizations. To analyze those four problems, four different techniques and combinations thereof were applied. The techniques used were linear and nonlinear programming, dynamic programming (DP), and the Lagrange multipliers method. Complete mixing of water was assumed to occur in the reservoirs in their study. In the model developed based on the conventional DP technique, the volume and salt concentration in the reservoir at the end of a particular time period were treated as two state variables.

Even the modeling of a conservative variable like salt concentration in the reservoir release is complex. The assumption of complete mixing of water in the reservoir throughout the year reduces the complexity involved with a stratified reservoir (which is the real case) to a certain extent. In such a problem, besides the continuity equation for water quantity, another continuity equation for salt exists. The continuity equation for the quantity and the continuity equation for the salt content of water in the reservoir are to be maintained. The salt concentration in the reservoir at each point in time is nonlinearly related to the volume and flow variables in the salt balance equation as given in Eq. (4.4).

The use of reservoir outlet works incorporating selective withdrawal structures is a primary method for controlling the quality of release. The optimum operation of these selective withdrawal structures is beneficial. To analyze such a problem, the DP technique may be applicable because of the sequential decision nature of the problem and the ability of DP to handle system nonlinearities conveniently.

Fontane et al. (1981) and Labadie and Fontane (1986) presented a technique for solving high-dimensional DP problems that condition optimal solutions on the one-dimensional objective space rather than the multidimensional state space. In this approach, a one-dimensional DP formulation in objective space replaces a high-dimensional DP problem involving the usual discretization of the state space.

They showed how the problem of determining optimal selective withdrawal structure operations could be solved over an objective space without the need to include the original state variables (vectors of average salt concentration and/or volume of layers of the reservoir) in the DP optimal value function. This, termed the objective-space DP approach, could reduce the original multidimensional problem to a one-dimensional DP problem.

Crawley and Dandy (1989) and Dandy and Crawley (1990, 1992) adopted an approach combining optimization and simulation techniques to derive operation policies for a system of reservoirs when water quality (salinity) was an important consideration in the system operation. They formulated a model that simulates salinity in a reservoir and it was run with an optimization model that considers only the quantity requirement, in an iterative fashion.

Their optimization (quantity) model is based on the linear programming technique. The quality model assumes complete mixing of water in the reservoir. This is a simplification compared with the real behavior of stratification occurring in reservoirs. The results indicated that improved operation policies with respect to reducing the cost involved (due to water of poor quality) could be derived from the methodology.

A number of studies have examined the use of reservoirs for control of downstream water quality (Jaworski et al., 1970; Orlob and Simonovic, 1981; Simonovic and Orlob, 1981, 1984). In all cases, water quality in the reservoir was not modeled. The regulation of streamflows needed to assimilate or dilute waste loads has been studied.

4.1 IDP BASED MODELS IN RESERVOIR OPERATION FOR QUALITY

This section presents two optimization models developed by Nandalal (1995) for optimum operation of a reservoir when both quantity and quality of water supplied from it are of interest. The models are based on the IDP technique. One model considers only releases while the other model considers both inflows and releases in the improvement of the quality of water supplied from the reservoir. Inflow manipulation is achieved by diverting (bypassing) inflows before they reach the body of the reservoir. Outflow manipulation includes release of excess water of (relatively) high salinity from the reservoir at appropriate times to flush (cleanse) the reservoir. Nonlinear salt balance constraints are included in both optimization models.

4.1.1 Optimization Model 1: controlling discharges only

In this case, only discharges from the reservoir can be manipulated. Complete mixing is assumed to occur in the reservoir.

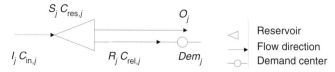

Figure 4.1 System configuration: Optimization Model 1

The system is as shown in Figure 4.1. The reservoir is operated on a monthly basis. Rates of inflow, outflow, and spill for the reservoir are constant during each time period. The forward algorithm of DP is used in the optimization procedure. The general scheme of the IDP procedure presented in Section 1.5 is used in the formulation of the model.

The objective function used in the model is to minimize the weighted summation of the squared deviation of release salinity and reservoir salinity from their respective target levels over the total period considered. Downstream quantity demand is treated as a constraint.

$$\text{OF} = \text{Minimize} \sum_{j=1}^{N} \left[W_1 (C_{\text{rel},j} - C_{\text{trg},j})^2 + W_2 \left(C_{\text{res},j+1} - \hat{C}_{\text{trg},j} \right)^2 \right],$$

$$(4.1)$$

where

$C_{\text{trg},j}$ = target release salinity during period j (ppm),
$\hat{C}_{\text{trg},j}$ = target reservoir salinity during period j (ppm),
$C_{\text{rel},j}$ = average salinity of release during period j (ppm),
$C_{\text{res},j+1}$ = average salinity of reservoir at end of period j (ppm),
N = number of periods,
W_1, W_2 = weightages, and
j = time period $(1, 2, \ldots, N)$.
Reservoir storage and release are assumed to be the state variable and the decision variable, respectively. The minimization is subject to constraints in storage volume, release, and conservation of salt.

STORAGE VOLUME CONSTRAINT
The storage volumes at the beginning of the first period and at the end of the last period are fixed. For all the other periods the volume belongs to the set of admissible storage volumes:

$$S_{\min} \leq S_{j+1} \leq S_{\max}; \qquad j = 1, 2, \ldots, N-1, \qquad (4.2)$$

where
S_{j+1} = storage volume at end of period j $(10^6 \, \text{m}^3)$, and
$S_{\max}$ and $S_{\min}$ = maximum and minimum storage volumes of reservoir $(10^6 \, \text{m}^3)$.

RELEASE CONSTRAINT
Maximum release from the reservoir is limited to the allowable release through the outlet. Minimum release is specified

by downstream irrigation demand, which is an implicit objective to be satisfied in the operation of the reservoir.

$$R_{\min,j} \le R_j \le R_{\max}; \qquad j = 1, 2, \ldots, N, \qquad (4.3)$$

where

R_j = release during period j ($10^6\,\mathrm{m}^3$),

$R_{\max}$ = maximum allowable release through outlet ($10^6\,\mathrm{m}^3$), and

$R_{\min,j}$ = irrigation demand during period j ($10^6\,\mathrm{m}^3$).

CONSERVATION OF SALT

The constraint that represents the conservation of salt in the reservoir is

$$S_{j+1}C_{\mathrm{res},j+1} = S_j C_{\mathrm{res},j} + I_j C_{\mathrm{in},j} - R_j C_{\mathrm{rel},j} - O_j C_{\mathrm{o},j}, \qquad (4.4)$$

where

$C_{\mathrm{in},j}$ = average salinity of inflow during period j (ppm),

$C_{\mathrm{o},j}$ = average salinity of spill during period j (ppm),

$C_{\mathrm{rel},j}$ = average salinity of release during period j (ppm),

$C_{\mathrm{res},j}$ = average reservoir salinity at beginning of period j (ppm),

I_j = inflow during period j ($10^6\,\mathrm{m}^3$), and

O_j = spill during period j ($10^6\,\mathrm{m}^3$).

Other variables are as defined before. Evaporation terms do not enter the salt balance as it is assumed that no salt is contained in the evaporating liquid.

The following equations are used to assess salinity in the reservoir at the end of period j. Derivation of these equations is described in Section 4.1.4.

If the reservoir volume is changing during period j,

$$C_{\mathrm{res},j+1} = \frac{1}{(Q_j+b)} \left[I_j C_{\mathrm{in},j} - [I_j C_{\mathrm{in},j} - C_{\mathrm{res},j}(Q_j+b)] \left(\frac{S_{j+1}}{S_j} \right)^{-(Q_j+b)/b} \right]; \qquad (4.5)$$

if the reservoir volume is constant during period j,

$$C_{\mathrm{res},j+1} = \frac{1}{Q_j} \left[I_j C_{\mathrm{in},j} - [I_j C_{\mathrm{in},j} - C_{\mathrm{res},j} Q_j] \exp\left(-\frac{Q_j}{S_j} \right) \right]; \qquad (4.6)$$

and the average salinity of spill during period j is

$$C_{\mathrm{o},j} = \frac{I_j C_{\mathrm{in},j}}{Q_j} + \frac{S_j}{Q_j^2} [I_j C_{\mathrm{in},j} - C_{\mathrm{res},j} Q_j] \left\{ \exp\left(-\frac{Q_j}{S_j} \right) - 1 \right\}; \qquad (4.7)$$

where

Q_j = total outflow (total of release and spill) during period j, and

b = change of reservoir storage during period j.

STATE TRANSFORMATION EQUATION

Based on the principle of continuity of the reservoir,

$$S_{j+1} = S_j + I_j - R_j - E_j - O_j, \qquad (4.8)$$

where E_j is the evaporation during period j ($10^6\,\mathrm{m}^3$). The other variables are as defined above.

RECURSIVE EQUATION

The DP recursive equation is formulated as

$$F_{j+1}^*(S_{j+1}) = \min_j \left\{ \mathrm{SQD}_j + F_j^*(S_j) \right\}, \qquad (4.9)$$

where

$F_{j+1}^*(S_{j+1})$ = minimum accumulated value of objective function from stage 0 to stage $j+1$, when state at stage $j+1$ is S_{j+1}, and

$\mathrm{SQD}_j = \left\lfloor W_1 \left(C_{\mathrm{rel},j} - \hat{C}_{\mathrm{trg},j} \right)^2 + W_2 \left(C_{\mathrm{res},j+1} - \hat{C}_{\mathrm{trg},j} \right)^2 \right\rfloor$

= weighted summation of squared deviation of release salinity and reservoir salinity from their respective target levels.

4.1.2 Optimization Model 2: controlling both inflows and discharges

Both inflows and releases can be controlled in the improvement of the quality of water supplied from reservoirs while satisfying quantity demand. The system is as shown in Figure 4.2. Provision to divert part of the inflow to a bypass whenever necessary has been introduced in this system. Complete mixing is assumed to occur in the reservoir throughout the year. The reservoir is operated on a monthly basis. Rates of inflow, outflow, diversion, and spill for the reservoir are constant during each time period. The forward algorithm of DP is used in the optimization procedure. The general scheme of the IDP procedure presented in Section 1.5 is used in the formulation of the model.

The objective function is the same as that in Optimization Model 1 (Eq. 4.1). Reservoir storage is the state variable while release and diversion are the decision variables. The minimization is subjected to storage volume and release constraints presented under Optimization Model 1 (Eq. 4.2 and Eq. 4.3). The diversion from the inflow is constrained by an allowable limit:

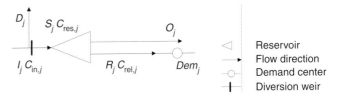

Figure 4.2 System configuration: Optimization Model 2

$$0 \le D_j \le D_{\max}; \qquad j = 1, 2, \ldots, N, \qquad (4.10)$$

where

$D_{\max}$ = maximum allowable diversion from inflow in a month ($10^6\,\mathrm{m}^3$), and

D_j = total diversion made during period j ($10^6\,\mathrm{m}^3$).

Diversion during a certain month is always less than or equal to the inflow in that month:

$$0 \le D_j \le I_j; \qquad j = 1, 2, \ldots, N. \qquad (4.11)$$

CONSERVATION OF SALT

The constraint that represents the conservation of salt in the reservoir is

$$S_{j+1} C_{\mathrm{res},j+1} = S_j C_{\mathrm{res},j} + (I_j - D_j) C_{\mathrm{in},j} - R_j C_{\mathrm{rel},j} - C_{\mathrm{o},j} O_j. \qquad (4.12)$$

Equations (4.5)–(4.7) are used to assess the salinity in the reservoir at the end of period j and the salinity of the spill during period j.

STATE TRANSFORMATION EQUATION

Based on the principle of continuity of the reservoir,

$$S_{j+1} = S_j + I_j - D_j - R_j - E_j - O_j. \qquad (4.13)$$

RECURSIVE EQUATION

The DP recursive equation is the same as Eq. (4.9) presented under Optimization Model 1.

4.1.3 Simulation model: completely mixed reservoir

A simulation model was formulated to simulate reservoir operation according to a prespecified release pattern described below. The reservoir is assumed to be completely mixed throughout the year. The simulation model uses Eqs. (4.4)–(4.8) in the regulation of the reservoir. Furthermore, the simulation procedure considers constraints for reservoir storages and releases as given in Eq. (4.2) and Eq. (4.3), respectively. This model furnishes end of month reservoir salinities and monthly average release salinities. The two optimization models are compared with results obtained from this simulation model.

In the release pattern adopted in the simulation model, the primary operation criterion is to make mandatory releases (downstream demands) only. However, if this criterion is strictly followed it is inevitable that the reservoir storage reaches maximum volume before the end of the period in certain months. If this happens, then the excess volume of water has to spill. In such instances the above policy to release only demand is over-ruled. The excess volume of water is released through the outlet subject to maximum allowable release. If it exceeds the maximum limit, the additional volume spills. The monthly demand is not totally satisfied only if there is not enough water in the reservoir. However, in such cases water available in the reservoir is supplied to satisfy the demand at least partly. This operation pattern is designated as "standard release policy."

4.1.4 Model of salinity in a reservoir

For a reservoir that is completely mixed, the continuity equation (salt balance equation) is

$$\frac{d(SC)}{dt} = IC_{\mathrm{in}} - QC, \qquad (4.14)$$

where

C = instantaneous salinity in reservoir at time t,

S = instantaneous volume of storage in reservoir at time t,

I = rate of total inflow,

C_{in} = average salinity of total inflow, and

Q = rate of total outflow including irrigation supply and spill; i.e.,

$$S\frac{dC}{dt} + C\frac{dS}{dt} = IC_{\mathrm{in}} - QC. \qquad (4.15)$$

If the rates of inflow and outflow are assumed to be constant, then

$$\frac{dS}{dt} = b \qquad (4.16)$$

and

$$S = a + bt, \qquad (4.17)$$

where a and b are constants.

Therefore, substituting Eq. (4.16) and Eq. (4.17) into Eq. (4.15) gives

$$(a + bt)\frac{dC}{dt} + Cb = IC_{\mathrm{in}} - QC. \qquad (4.18)$$

By rearranging:

$$\frac{dC}{dt} = \frac{IC_{\mathrm{in}} - QC - bC}{a + bt} = \frac{IC_{\mathrm{in}} - C(Q + b)}{a + bt}. \qquad (4.19)$$

Rearranging and integrating the above equation from time t_j to t_{j+1}:

The salinity in the reservoir changes from $C_{\mathrm{res},j}$ to $C_{\mathrm{res},j+1}$ while the storage changes from S_j to S_{j+1}.

$$\int_{C_{\text{res},j}}^{C_{\text{res},j+1}} \frac{dC}{IC_{\text{in}} - C(Q + b)} = \int_{t_j}^{t_{j+1}} \frac{dt}{(a + bt)},$$

$$\frac{1}{(Q + b)} \int_{C_{\text{res},j}}^{C_{\text{res},j+1}} \frac{dC}{[IC_{\text{in}}/(Q + b) - C]} = \frac{1}{b} \int_{t_j}^{t_{j+1}} \frac{dt}{(a/b + t)},$$

$$\int_{C_{\text{res},j}}^{C_{\text{res},j+1}} \frac{dC}{[IC_{\text{in}}/(Q + b) - C]} = \frac{(Q + b)}{b} \int_{t_j}^{t_{j+1}} \frac{dt}{(a/b + t)},$$

$$-\ln\left[\frac{IC_{\text{in}}}{(Q + b)} - C\right]_{C_{\text{res},j}}^{C_{\text{res},j+1}} = \frac{(Q + b)}{b} \ln\left(\frac{a}{b} + t\right)_{t_j}^{t_{j+1}},$$

$$\ln\left[\frac{IC_{\text{in}}/(Q + b) - C_{\text{res},j+1}}{IC_{\text{in}}/(Q + b) - C_{\text{res},j}}\right] = -\frac{(Q + b)}{b} \ln\left(\frac{a/b + t_{j+1}}{a/b + t_j}\right),$$

$$\frac{IC_{\text{in}} - C_{\text{res},j+1}(Q + b)}{IC_{\text{in}} - C_{\text{res},j}(Q + b)} = \left(\frac{a + bt_{j+1}}{a + bt_j}\right)^{-(Q+b)/b} = \left(\frac{S_{j+1}}{S_j}\right)^{-(Q+b)/b},$$

$$IC_{\text{in}} - C_{\text{res},j+1}(Q + b) = [IC_{\text{in}} - C_{\text{res},j}(Q + b)]\left(\frac{S_{j+1}}{S_j}\right)^{-(Q+b)/b}.$$

At the end of the time period the salinity in the reservoir is

$$C_{\text{res},j+1} = \frac{1}{(Q + b)}\left[IC_{\text{in}} - [IC_{\text{in}} - C_{\text{res},j}(Q + b)]\left(\frac{S_{j+1}}{S_j}\right)^{-(Q+b)/b}\right]$$

For the special case where the volume of the reservoir is not changing (i.e., $b = 0$):

The continuity equation, Eq. (4.14),

$$\frac{d(SC)}{dt} = IC_{\text{in}} - QC,$$

i.e.,

$$S\frac{dC}{dt} + C\frac{dS}{dt} = IC_{\text{in}} - QC.$$

If the storage is constant,

$$S = A \ (= \text{a constant}) \rightarrow \frac{dS}{dt} = 0,$$

$$A\frac{dC}{dt} = IC_{\text{in}} - QC.$$

By rearranging,

$$\frac{dC}{(IC_{\text{in}}/Q - C)} = \frac{Q}{A} dt.$$

Integrating the above equation from time t_j to t_{j+1} with the change in salinity from $C_{\text{res},j}$ to $C_{\text{res},j+1}$:

$$\int_{C_{\text{res},j}}^{C_{\text{res},j+1}} \frac{dC}{(IC_{\text{in}}/Q - C)} = \frac{Q}{A} \int_{t_j}^{t_{j+1}} dt,$$

$$-\ln(IC_{\text{in}}/Q - C)\Big|_{C_{\text{res},j}}^{C_{\text{res},j+1}} = \frac{Q}{A}(t_{j+1} - t_j).$$

If $t_{j+1} - t_j = \Delta t$, then:

$$\ln\left(\frac{IC_{\text{in}} - C_{\text{res},j+1}Q}{IC_{\text{in}} - C_{\text{res},j}Q}\right) = -\frac{Q\Delta t}{A},$$

$$\frac{IC_{\text{in}} - C_{\text{res},j+1}Q}{IC_{\text{in}} - C_{\text{res},j}Q} = \exp\left(-\frac{Q\Delta t}{A}\right),$$

$$IC_{\text{in}} - C_{\text{res},j+1}Q = [IC_{\text{in}} - C_{\text{res},j}Q]\exp\left(-\frac{Q\Delta t}{S_j}\right).$$

At the end of the time period the salinity in the reservoir is

$$C_{\text{res},j+1} = \frac{1}{Q}\left[IC_{\text{in}} - [IC_{\text{in}} - C_{\text{res},j}Q]\exp\left(-\frac{Q\Delta t}{S_j}\right)\right]. \quad (4.20)$$

SPILLED SALT LOAD

During spill there is a constant volume of water in the reservoir (ignoring the effect of surcharge). Therefore, Eq. (4.20) can be used to determine the spilled salt load and hence the average salinity of the spilled water.

Let

L_s = spilled salt load during the time period,

O_s = volume of spill per unit time during the time period (assumed constant), and

C_o = average salinity of spill during the time period.

Then

$$L_s = \int_{t_j}^{t_{j+1}} O_s C\, dt$$

$$= \int_{t_j}^{t_{j+1}} (O_s/Q)\left[IC_{\text{in}} - [IC_{\text{in}} - C_{\text{res},j}Q]\exp\left(-\frac{Q\Delta t}{S_j}\right)\right]dt$$

$$= \frac{O_s}{Q}IC_{\text{in}}\Delta t + \frac{O_s S_j}{Q^2}[IC_{\text{in}} - C_{\text{res},j}Q]\left[\exp\left(-\frac{Q\Delta t}{S_j}\right) - 1\right], \quad (4.21)$$

but $C_o = L_s/(O_s\Delta t)$

$$C_o = \frac{IC_{\text{in}}}{Q} + \frac{S_j}{Q^2\Delta t}[IC_{\text{in}} - C_{\text{res},j}Q]\left[\exp\left(-\frac{Q\Delta t}{S_j}\right) - 1\right]. \quad (4.22)$$

4.2 THE JARREH RESERVOIR IN IRAN

The Jarreh Reservoir, built to irrigate 13 000 ha, is one of the projects coming under the water resources development plan for the Shapur–Dalaki River basin in Iran. The Shapur–Dalaki River basin is located in southwest Iran (long. 52° 20′, 50° 45′ E; lat. 30° 02′, 28° 45′ N) as shown in Figure 4.3. The uplands of the basin are mountainous with a maximum elevation of 3000 m above MSL. The altitude

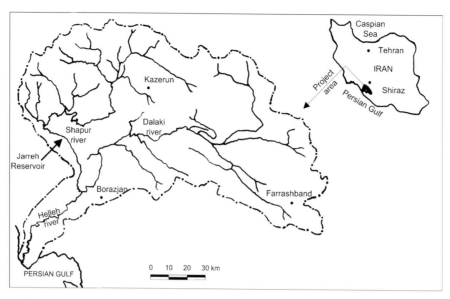

Figure 4.3 The Shapur–Dalaki basin

decreases to about 20 m at the confluence of the Shapur and Dalaki Rivers on the coastal plain. Total drainage area is approximately 10 000 km², of which the Shapur River and its tributaries drain 4110 km² of the northern region and the Dalaki River and its tributaries drain 5800 km² of the southern region. The rivers join to form the Helleh River, which debouches into the Persian Gulf.

The climate of the Shapur–Dalaki basin is classified as arid (Shiati, 1991); average annual rainfall is below 20 percent of total annual potential evaporation. The degree of aridity is lower only in higher parts of the basin, where precipitation is relatively high.

Except in occasional wet years, most precipitation is confined to the winter months in this basin. The dry season lasts from April to October. Total annual rainfall decreases southwards towards the coastal plains and the Persian Gulf. Mean value varies between 600 mm in the upper part of the basin to less than 200 mm along the coast. Rainfall occurs mainly during the six months of November to April with a peak in midwinter.

In the basin, a great variation of mean temperature is observed over the year. Mean annual values range between 16 °C in the highest (northern) part of the basin and 24 °C in the southwestern coastal plain. The maximum and minimum temperatures occur in July/August and January/February, respectively. Frosts are common in the interior, but rare on the coastal plain. Daily variation in temperature is very high in all parts of the basin. Annual potential water evaporation is high and its mean value is about 2000 mm. Mean yearly relative humidity in the area is around 55%, and follows a clear seasonal trend.

Average annual flows in the Shapur and Dalaki Rivers are about 530×10^6 m³/year and 425×10^6 m³/year respectively. Variation of flows from year to year is considerable. The discharge mainly occurs during winter, and reaches a maximum in February.

The Shapur and Dalaki Rivers primarily originate from karstic springs, which yield water of excellent quality. Further downstream they pass through large areas with salt domes and saline erodible formations. As a consequence, they become increasingly contaminated by salts. Severe erosion of these scarcely vegetated and soft materials results in very high salt content of this water.

The high salt content of the water forms an obstacle to its use for irrigation. Average total dissolved solids in the Shapur River, where the Jarreh Reservoir is built are 3700–4000 ppm in summer and 2130–2444 ppm in winter.

Agriculture has been practised for centuries in this basin. The inland basins filled with fertile alluvial soils and parts of the coastal plain are intensively cultivated. The steep hills and mountains and the saline parts of the coastal plains are used for grazing, mainly with sheep.

However, shortage of water, salinity of water, and adverse chemical and physical soil properties are found to impede the agricultural development of the alluvial plains. According to Yekom Consulting Engineers (1980), out of 86 000 ha of irrigable lands, about 46 000 ha could be irrigated through implementation of several water resources development projects within this basin. The high salinity – of sodium chloride type – only allows farmers to grow crops such as date palm, barley, wheat, and alfalfa that have sufficient tolerances to salinity.

Table 4.1. *Monthly irrigation demands (for 13 000 ha)*

Month	Jan.	Feb.	Mar.	Apr.	May.	Jun.	Jul.	Aug.	Sep.	Oct.	Nov.	Dec.
Irrigation demand ($10^6 \, m^3$/month)	17.5	23.0	34.0	26.5	19.5	22.0	26.5	30.0	27.0	22.0	10.5	12.0

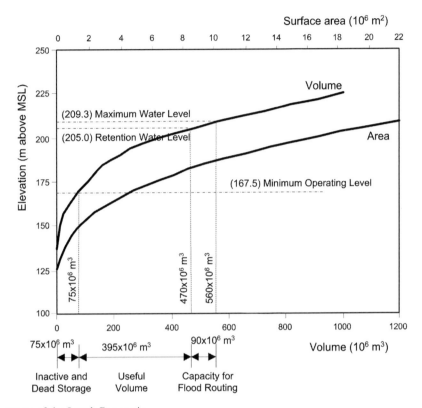

Figure 4.4 Characteristic curves of the Jarreh Reservoir

The Shapur and Dalaki Rivers possess a regime of flash floods in winter. During summer drought, their flow falls to very low values. Therefore, only a storage dam could regulate the flow of the river needed to create the conditions necessary for developing agricultural resources. In addition, salinity can be regulated and improved by careful management of such reservoirs.

The Jarreh storage dam to irrigate about 13 000 ha is one of the projects. The location of this reservoir is shown in Figure 4.3. The behavior of the salt-affected Jarreh Reservoir is of great concern. Catchment management measures to reduce salinity are less effective in this sparsely vegetated basin since the existence of salt formations hampers plant growth. Shiati (1991) showed that the Jarreh Reservoir could regulate and reduce the salt concentration of the irrigation water to a range between 1500 and 2400 ppm from a range between 900 and 4000 ppm. Careful management of the reservoir may further improve the quality of the water released. This

chapter presents a comprehensive study carried out on the operation of the Jarreh Reservoir for the improvement of water quality.

The irrigation demands (Shiati, 1991) to be supplied from the Jarreh Reservoir are given in Table 4.1. Effective storage–surface area–elevation relationships of the Jarreh Reservoir are shown in Figure 4.4. The salient features of the dam and the reservoir are summarized in Table 4.2.

4.3 APPLICATION OF THE MODELS TO THE JARREH RESERVOIR

Discharge and salinity of the Shapur River at the Jarreh Reservoir dam site are shown for the period 1975–89 in Figure 4.5. As it shows, salinity of the river is strongly influenced by long-term variations in streamflow in addition to its seasonal variations. For example, the high annual discharge

Table 4.2. *Salient features of the Jarreh dam and reservoir*

Reservoir	
Normal high water level (retention level)	205.0 m MSL
Normal storage capacity	470.0 $(10^6\,\mathrm{m}^3)$
Minimum water surface level	167.5 m MSL
Minimum storage capacity (dead storage)	75.0 $(10^6\,\mathrm{m}^3)$
Maximum flood level	209.3 m MSL
Water surface area (at normal retention level)	19.5 km^2

Dam		
Type	Concrete arch dam (double curvature)	
Elevation at crest		210.5 m MSL
Length at crest		215.0 m
Minimum thickness of dam (at elevation 205.0 m MSL)		3.0 m
Maximum thickness of dam (at elevation 125.0 m MSL)		11.0 m

Spillway		
Number of spillways: 3 (two morning glory and one overflow spillway)		
Morning glory spillway (right bank)	Maximum capacity	1200.0 m^3/s
	Sill elevation	205.0 m MSL
Morning glory spillway (left bank)	Maximum capacity	1200.0 m^3/s
	Sill elevation	205.0 m MSL
Overflow spillway	Maximum capacity	1650.0 m^3/s
	Sill elevation	205.0 m MSL
Total discharge at 209.3 m MSL		4400.0 m^3/s

Source: Yekom Consulting Engineers (1980)

of 1982 (discharge $1127 \times 10^6\,\mathrm{m}^3$, 2.12 times the median) averaged 1500 ppm whereas the discharge of 1984, a dry year (discharge $280 \times 10^6\,\mathrm{m}^3$, 0.54 times median), averaged 2180 ppm in salinity. Consequently, flood flows (winter period) are generally less saline and low flows (summer period) remain highly saline. For the recorded period the salinity of the river water varies between 750 and 4200 ppm.

The Shapur River is characterized by high variability of both discharges and salinity. In wet years when the river flows are high, average salinity of inflow is low and the reservoir could be flushed out improving the quality of impounded water. On the other hand, a dry year causes a considerable deterioration of quality, and a series of consecutive dry years will deteriorate the quality even more.

4.3.1 Optimization Model 1: controlling discharges only

The Jarreh Reservoir has a total storage capacity of $470 \times 10^6\,\mathrm{m}^3$ and a dead storage capacity of $75 \times 10^6\,\mathrm{m}^3$. The allowable release through the reservoir is limited to $150 \times 10^6\,\mathrm{m}^3$ in a month. The target release and reservoir salinities are set to 1000 ppm. The release and reservoir salinities were always higher than this value. The monthly irrigation demands are

given in Table 4.1. The period considered in a single optimization is 15 years, the total number of stages being 180 (12 months × 15 years).

COMPARISON OF COMPONENTS IN THE OBJECTIVE FUNCTION

The objective function used in the model (see Eq. 4.1) has two parts. They are:

(a) to minimize the deviation of release salinity from a target; and

(b) to minimize the deviation of reservoir salinity from a target.

Different weights given to the two components show their impact on the final aim of reducing release salinity.

(a) $W_1 = 1.0$, $W_2 = 0.0$, only release salinity is considered.

(b) $W_1 = 0.5$, $W_2 = 0.5$, both release salinity and reservoir salinity are considered with equal importance.

(c) $W_1 = 0.0$, $W_2 = 1.0$, only reservoir salinity is considered.

Table 4.3 presents the monthly average release salinities obtained based on the IDP model.

Table 4.3. *Comparison of different objective functions*

Weights		OF value/10^6	Total release (10^6 m^3)	Total spill (10^6 m^3)	Monthly average salinity (ppm)	
W_1	W_2				Reservoir	Release
1.0	0.0	134.5	8043	89	1832	1828
0.5	0.5	132.2	8090	47	1820	1815
0.0	1.0	133.4	8089	46	1821	1816

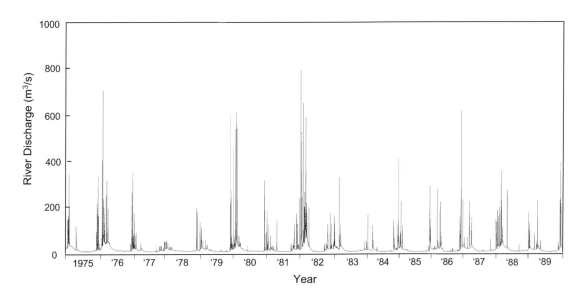

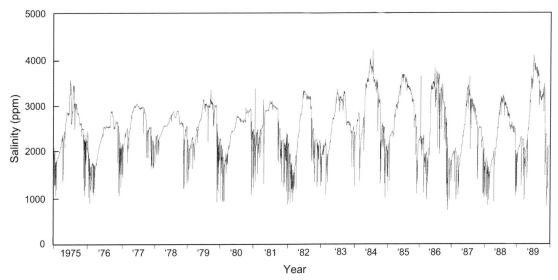

Figure 4.5 River discharges and salinities: 1975–89

Differences observed among the above three objective functions are almost negligible. Thus, the improvements obtainable in quality of the release water, by considering the quality of either the release or the reservoir, or both, in the objective function, are similar. The second alternative that gives equal weight to the two components is used in the subsequent analysis.

Release salinities obtained from Optimization Model 1 were compared with that obtained from a reservoir operation

Table 4.4. *Comparison of IDP optimum operation with simulation*

Operation	OF value/10^6	Total release (includes scour) ($10^6\,\mathrm{m}^3$)	Total spill ($10^6\,\mathrm{m}^3$)	Monthly average salinity (ppm)	
				Reservoir	Release
Optimization Model 1	132.2	8090	47	1820	1815
Simulation (Standard release policy)	166.2	7263	729	1939	1939

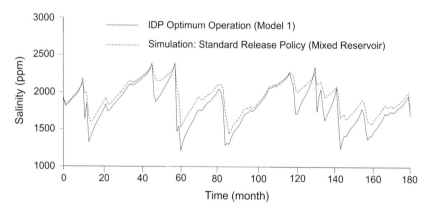

Figure 4.6 Reservoir salinity: comparison of IDP optimum operation with standard release policy

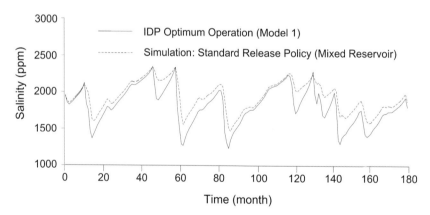

Figure 4.7 Monthly average release salinity: comparison of IDP optimum operation with standard release policy

simulation model, which adopts "standard release policy" and assumes complete mixing of water throughout the year.

Table 4.4, which compares the two operations, indicates that the IDP optimum operation results in an improved operation pattern. This is the best release pattern for improving quality of water when only release could be manipulated. Note that the optimization model runs with perfect knowledge of inflows. Spill has been reduced in the IDP based optimization compared with the simulation. This reduced

spill has been used to flush the reservoir whenever possible, thereby improving the quality of water in both the reservoir and the releases. The increase in release (that includes scour/flush volume) indicates this.

The end of month reservoir salinities and monthly average release salinities for these two cases are compared in Figure 4.6 and Figure 4.7, respectively. These figures indicate that the IDP based optimum operation is superior to simulation throughout the total period of 15 years.

Table 4.5. *Releases of IDP optimization*

Month	Average monthly inflow ($10^6 \, m^3$)	Average monthly release ($10^6 \, m^3$)	Demand ($10^6 \, m^3$)	Average of additional releases ($10^6 \, m^3$)
January	90.75	50.21	17.50	32.71
February	105.41	37.42	23.00	14.42
March	91.23	52.72	34.00	18.72
April	57.44	32.51	26.50	6.01
May	25.77	23.55	19.50	4.05
June	13.33	22.22	22.00	0.77
July	9.76	26.55	26.50	0.05
August	10.55	30.42	30.00	0.42
September	10.85	46.36	27.00	19.36
October	17.28	63.23	22.00	41.23
November	30.26	89.11	10.50	78.11
December	97.63	64.52	12.00	52.52

Average monthly inflows and average monthly releases (obtained from the IDP model) are compared with demands in Table 4.5. Additional releases represent the amount of water released beyond the downstream demand. This additional volume of water is used for flushing (or scouring) the reservoir. Table 4.5 indicates that the flushing of the reservoir occurs mainly in autumn and early winter (Sep.–Dec.) when the quality of water in the reservoir is poor (Sep.–Nov.). This is followed by a significant improvement in the quality of water in the reservoir due to high inflows of good quality in winter and early spring (Dec.–Mar.). Although flushing continues through winter till early spring, the quantity is less compared with that in autumn.

IMPACT OF QUALITY IN THE
OBJECTIVE FUNCTION
Comparison of release salinity obtained from the IDP model with release salinity obtained from an optimization model that considers only downstream quantity requirements is presented next. This comparison is designed to examine the effectiveness of the inclusion of quality considerations into the optimization model. A modified version of Optimization Model 1 considering only the downstream quantity demand was used at this step. The objective function is to minimize the squared deviation of release from the demand over the total period; i.e.,

$$OF = \text{Minimize} \sum_{j=1}^{N} (R_j - \text{Dem}_j)^2. \qquad (4.23)$$

Optimization was carried out with the same set of inflow data (15 years, from 1974 to 1989). The model has the same storage volume constraints (Eq. 4.2) as in Optimization

Model 1. But release is only limited by the maximum allowable amount in Eq. (4.3). The state transformation equation is the same as in Optimization Model 1 (Eq. 4.8). The results presented in Table 4.6 and Figure 4.8 show the effectiveness of including quality considerations in the optimization on the improvement of quality.

Inclusion of quality in the model changes the operation policy. According to the changed policy, increased releases in autumn and early winter (to flush the reservoir) and reduced releases in the summer are observed.

4.3.2 Optimization Model 2: controlling both inflows and discharges

Optimization Model 2 controls both inflows and outflows in the improvement of supply water quality. It was run employing the same set of data and other parameters used in Optimization Model 1.

EFFECT OF ALLOWABLE MAXIMUM
DIVERSION
An important feature of the model is the ability to divert inflows (or bypass inflows) in addition to manipulation of releases. However, the maximum quantity of water that could be diverted in a month may be limited due to practical considerations such as capacity of diversion structures, canals, etc. The influence of the diversion limit on the improvement of supply water quality obtained based on the model is presented in Table 4.7 for several allowable diversion limits.

Release salinity and reservoir salinity improve with an increase of the allowable diversion limit. This is due to the increase in total volume of water diverted and total salt load

Table 4.6. *Comparison of two optimizations: effect of inclusion of quality*

IDP model (objective function)	Total release (includes scour) ($10^6\,m^3$)	Total spill ($10^6\,m^3$)	Average salinity (ppm)	
			Reservoir	Release
Quantity only	7438	684	1915	1913
Quantity and quality	8090	47	1820	1815

Table 4.7. *Effect of allowable maximum diversion*

Allowable diversion $10^6\,m^3$/month	OF value/10^6	Total release ($10^6\,m^3$)	Total spill ($10^6\,m^3$)	Diversion		Mean salinity (ppm)	
				Volume ($10^6\,m^3$)	Salt load $10^6\,kg$	Reservoir	Release
10	97.7	6793	47	1315	3356	1706	1703
20	79.8	5929	47	2189	5302	1638	1636
30	74.0	5506	47	2617	6134	1611	1608
40	71.0	5348	47	2780	6442	1595	1593
50	69.0	5226	47	2899	6670	1590	1586
60	68.0	5166	47	2972	6807	1578	1576
70	67.5	5141	47	2997	6847	1576	1574
80	67.4	5132	47	3006	6853	1575	1573

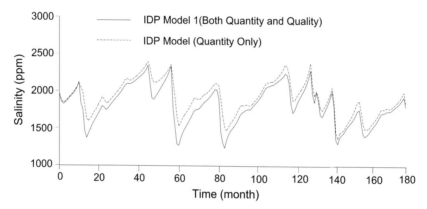

Figure 4.8 Monthly average release salinity: effect of including quality considerations in the optimization model

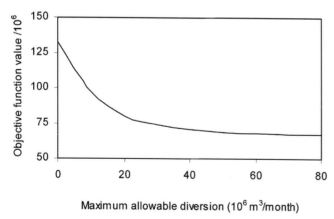

Figure 4.9 Objective function value for different allowable diversion limits

diverted. Note that the total release has been decreased. This implies that the diversion of poor quality inflows has a strong influence on the improvement of quality in the reservoir and the releases.

But the improvements diminish with an increase in the diversion limit as shown in Figure 4.9. For the Jarreh Reservoir, increasing the allowable diversion limit above $40 \times 10^6\,m^3$/month is not influential in reducing the concentrations of release or reservoir salinity significantly.

The analysis of diversions shows that when the allowable diversion limit is $80 \times 10^6\,m^3$/month, the maximum diversion observed was only $79.5 \times 10^6\,m^3$/month. Thus, a further increase in the allowable diversion limit would not be effective

Table 4.8. *Comparison of optimum diversions with cut-off level diversions*

Alternative	OF value 10^6	Total release (10^6 m^3)	Total spill (10^6 m^3)	Total diversion (10^6 m^3)	Mean salinity (ppm)	
					Reservoir	Release
Cut-off at 3000 ppm	147.7	7000	703	297	1884	1884
Cut-off at 2800 ppm	135.0	6740	702	564	1846	1847
Cut-off at 2500 ppm	107.2	6166	689	1194	1756	1756
Model 2: max. diversion $10 \times 10^6 \text{ m}^3$/month	97.7	6793	47	1315	1706	1703
Model 2: max. diversion $80 \times 10^6 \text{ m}^3$/month	67.4	5132	47	3006	1575	1573

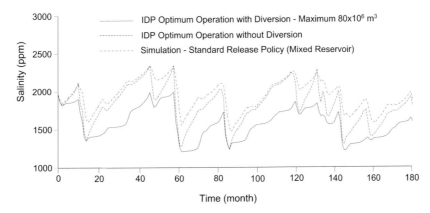

Figure 4.10 Monthly average release salinity: comparison of models

in improving quality for this set of data. However, from these observations it is apparent that the limitation on the allowable diversion affects the reductions in the reservoir and release salinities.

An analysis of release from the reservoir for different diversion limits indicates that most of the time optimum releases from the reservoir are the same. But the number of times the reservoir is flushed (cleansed) by larger winter flows is greater when the diversion limit is low compared with when the diversion limit is high. The model attempts to improve the quality of water by flushing the reservoir more, when the diversion is more restricted. However, this is less effective than diverting poor quality inflows. The analysis shows that inflow is mostly diverted during the summer low-flow period during which water quality is poor.

The monthly average release salinity distributions obtained from the two optimization models and the operation simulation with the standard release policy are compared in Figure 4.10. In Optimization Model 2 the allowable diversion was limited to $80 \times 10^6 \text{ m}^3$/month.

Release salinities obtained from the IDP optimum operation with diversions are observed to be the lowest throughout the total period. The above results suggest that the diversion of part of the inflows before entering the reservoir is the best management option for reducing the salinity level in the releases.

COMPARISON OF OPTIMUM DIVERSIONS WITH CUT-OFF DIVERSIONS

Bypassing inflows of poor quality above a prespecified (cut-off) level are effective in improving the quality of water released from a reservoir. Table 4.8 and Figure 4.11 compare the optimum operation obtained from Optimization Model 2 with standard release policy based operation simulation incorporating bypassing inflows having salinity above several cut-off levels.

The total quantity of diversions made in Optimization Model 2 with maximum diversion constrained to $10 \times 10^6 \text{ m}^3$/month is close to that in the simulation with cut-off at 2500 ppm. But the mean reservoir and release salinities obtained from Model 2 are observed to be less. About 32% of the time the diversions were more than $10 \times 10^6 \text{ m}^3$/month in the simulation with cut-off level at 2500 ppm. In certain months diversion quantity even rose to $22 \times 10^6 \text{ m}^3$/month. This requires larger diversion structures and canals, etc.

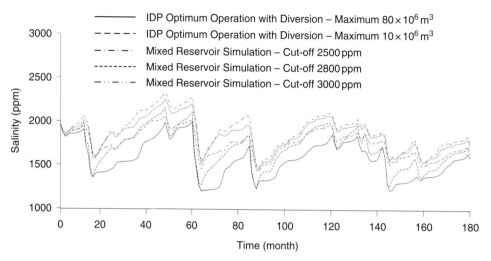

Figure 4.11 Monthly average release salinity: comparison of cut-off level with Optimization Model 2

Therefore, the IDP based (Model 2) optimum operation with allowable diversion limited to $10 \times 10^6\,\mathrm{m}^3/\mathrm{month}$ is superior to diverting inflows having salinity concentration above a cut-off level of 2500 ppm.

Among the alternatives compared in Table 4.8, the improvements obtained from Model 2 with allowable diversion limited to $80 \times 10^6\,\mathrm{m}^3/\mathrm{month}$ are apparently the best. Even though the total number of diversions is more in this operation, it does not risk violating the satisfaction of down-

stream quantity demand, since downstream quantity demand is treated as a constraint in the optimization model. Scrutiny of the results indicated that most of the diversions are in the summer when the inflows are of poor quality. Flushing the reservoir was observed to take place in winter when the inflows are substantial. However, note that it is not worth increasing the diversion capacity above $40 \times 10^6\,\mathrm{m}^3/\mathrm{month}$ as presented in Figure 4.9. The improvements obtained in performance hardly warrant increasing the diversion capacity twofold.

5 Large-scale reservoir system operation

5.1 USE OF DYNAMIC PROGRAMMING IN MULTIPLE-RESERVOIR OPERATION

In general, straightforward optimization of multiple-reservoir system operation with DP and SDP does not achieve the accuracy level relevant for real-world applications. The analysis of multiple-reservoir system operation imposes significant dimensionality problems due to the inevitable introduction of three inherent computational difficulties:

(a) Increase in dimensionality of the problem is reflected in the number of state and decision variables necessary to describe a multiple-reservoir system and its operation.
(b) Operation of complex reservoir systems involves multiple, and often noncommensurate objectives. Often these cannot be approximated by a single, clearly defined surrogate objective or criterion.
(c) Additional difficulties arise when it is necessary to consider stochasticity, which is an inherent feature in the operation of reservoir systems. Although it is common to reduce this aspect to river flow uncertainty only, the problem does not seem to be significantly alleviated because multiple reservoirs imply consideration of multiple, independent or cross-correlated, stochastic inflow processes.

Although regarded as a very promising stochastic optimization technique, SDP is still hampered by well-known dimensionality restrictions and the resulting huge computational requirements imposed when applied to multiple-state–multiple-decision problems. In general, and this was also reported by Yeh (1985), systems analysts have opted for one of the following three remedies, or combinations thereof, to overcome these difficulties:

(a) Decomposition of the system into smaller and simpler subsystems thus reducing the complex problem to a set of tractable tasks (e.g., decomposition based on physical or functional structure of the system, multilevel hierarchical decomposition, etc.) (Bogardi and Milutin, 1995; Bogardi et al., 1995; Ampitiya et al., 1996).
(b) Aggregation of the system, or parts thereof, into a composite system thus allowing a straightforward application of the optimization procedure and the subsequent disaggregation of the derived composite operating strategy into control policies of individual system elements (Kularathna and Bogardi, 1990; Kularathna, 1992).
(c) Replacement of discrete state, decision or objective function domains by their continuous approximations and subsequent application of complex mathematical methods to derive the optimal solution.

This section presents both deterministic and stochastic DP based applications in operational analyses of complex reservoir systems. Note that the deterministic optimization models present an integral part of implicit stochastic optimization techniques.

5.1.1 Decomposition based methodologies

Various decomposition approaches seem to be the most frequent means used to alleviate dimensionality problems in operational analysis of large-scale systems. Yeh (1985), for instance, observed that the majority of methods devised for dimensionality reduction involved some type of decomposition of the system into smaller and simpler subsystems, and the subsequent use of iterative procedures to find a solution to the complex problem. The advantage of decomposition is that it allows a large, unsolvable for a straightforward approach, problem to be reduced to a series of small tractable tasks. Furthermore, unlike continuous function approximation techniques, decomposition methods usually employ less complicated mathematical theories and, perhaps their most important characteristic, their computational complexity increases at a lower rate with the number of decomposed system elements. In general, decomposition based optimization approaches reach a local rather than the global optimum. Nevertheless, numerous studies have shown that near-optimal solutions derived by decomposition techniques could provide significant improvements in the operation of the systems in question. Thus, this technique has the potential for real-world use.

Heidari *et al.* (1971) introduced discrete differential dynamic programming (DDDP) to solve the deterministic optimization problem of a four-reservoir system. In essence, DDDP could be understood as an extension of IDP (Larson, 1968) to a multidimensional problem. Chow *et al.* (1975) analyzed the computer time and memory requirements for classical DP and DDDP and proposed the methodologies to estimate them. With regard to the necessary computer storage, they concluded that DDDP required substantially less data space than DP. Although a significant reduction in state and decision space size was evident, DDDP retained exponential growth in the number of system transitions with respect to the number of state variables. Aware of the fact, Nopmongcol and Askew (1976) proposed a decomposition approach named multilevel incremental dynamic programming (MIDP) to solve the same problem. The search for the optimal operational strategy of a multiple-reservoir system was carried out through several stages denoted as "one-at-a-time", "two-at-a-time", "three-at-a-time", etc. The core of the approach was that, at each stage, a set of individual IDP problems was solved, each of them having one, two, and three reservoirs taken into consideration, respectively. The search at each stage included all possible combinations of reservoirs (i.e., all single units, all pairs of reservoirs, all triplets, etc.). The procedure was terminated when no improvement of the objective function was observed at two consecutive levels. The convergence to the same result obtained by Heidari *et al.* (1971) was already observed after the second MIDP level (i.e., "two-at-a-time").

Trott and Yeh (1973) proposed a method to resolve the dimensionality problems inherent in operational analyses of multiunit reservoir systems with both serial and parallel connections. The sample system consisted of six water supply reservoirs with a single demand point located immediately below the lowest reservoir. The objective was to maximize the firm water supply at the demand location. They applied Bellman's method of successive approximations (Bellman, 1957; Bellman and Dreyfus, 1962) and used IDP to solve the decomposed, one-dimensional problems. The deterministic problem having six state variables was broken down into six problems having only one state variable and five equality constraints each. Thus, while optimizing the operation of a chosen reservoir, the operation policies of the remaining five reservoirs were kept constant as derived beforehand. Iterative cycles comprising six IDP runs were repeated until a stable benefit was observed in consecutive iterations. The prerequisite for starting the iterative procedure was to first select six independent and feasible operating strategies for each reservoir. The method was tested with three different sets of initial operating strategies. All trials converged towards the respective benefits falling within the 0.05% range of each other.

Turgeon (1980) applied two iterative decomposition techniques to optimize the long-term operation of a multiple-reservoir hydropower system consisting of a number of independent rivers, each with one or more serially connected reservoirs. Both approaches assumed that river flows were uncorrelated random processes. The first decomposition technique, named "one-at-a-time", broke down a system into a set of single-reservoir subsystems whose operations were optimized by SDP. The second, the "aggregation/decomposition" method, split up an *n*-reservoir system into *n* subsystems having two elements each. One of the elements corresponded to a selected single reservoir while the second described the hypothetical reservoir created by aggregating the remaining $n-1$ reservoirs into a single unit. Thus, the SDP optimization was in this case applied to a two-reservoir operation problem. The application of the former approach resulted in a local optimal operating strategy for each power plant, whereas the latter derived the global suboptimal operation policies for *n* individual reservoirs. The two models were compared on a pilot six-reservoir system and the "aggregation/decomposition" model derived better system returns in terms of the operating costs accumulated over the simulation period.

As a supplement to the previous work, Turgeon (1981) proposed an algorithm to derive monthly operating strategies for a hydropower system consisting of multiple, serially linked, reservoirs. The optimization was based on SDP considering monthly inflows to reservoirs as independent random processes. Basically, the approach decomposed an *n*-reservoir system into $n-1$ subsystems having two elements each. The elements of an *i*th subproblem were reservoir *i* and the respective hypothetical reservoir generated from all the remaining reservoirs situated downstream of *i* (i.e., reservoirs $i+1$, $i+2$, ... , *n*). The suboptimal operating strategy of reservoir *i* for a particular month defined the release policy as a function of its storage and the total amount of energy available in all downstream reservoirs. The main advantages of the algorithm were said to be the fact that it was not an iterative procedure, and that the computational requirements increased only linearly with the number of reservoirs.

To optimize the operation of a multiple-reservoir hydropower system Archibald *et al.* (1997) exploited a similar aggregation/decomposition idea. The method was devised to be applicable to any connected and acyclic reservoir network provided that the water released from any reservoir in the system directly and instantaneously enters at most one other reservoir. Consequently, the operating strategy for a reservoir could be determined by an SDP based model formulated for that reservoir and a two-dimensional representation of the rest of the system. Namely, given a particular reservoir, the remaining part of the system could be divided into a subset of reservoirs whose releases can reach the selected one, and a subset of the remainder of the system. To reduce the dimensionality of the optimization problem, aggregation was used to represent each of the subsets by a single hypothetical reservoir. The approach was tested on several reservoir systems, the largest containing 17 reservoirs. To evaluate their method, the authors derived the true optima for smaller test cases (i.e., having three and four reservoirs) by applying the equivalent full optimization models. In addition to substantial savings in processing time, the proposed decomposition/aggregation method was found to provide solutions close to the real optima.

Tai and Goulter (1987) developed an iterative algorithm for the optimization of a Y-shaped three-reservoir hydropower system (i.e., two parallel upstream reservoirs were serially connected to the third reservoir situated downstream). Only the downstream reservoir had hydroelectric generation facilities and the two upstream reservoirs served as storage regulation structures for the downstream one. The core of the method was a single-reservoir SDP based optimization model given in Loucks et al. (1981). Monthly inflows were assumed to be serially correlated and the first order Markov chain was used to describe transition probabilities between different inflow classes in consecutive months. Prior to starting the iterative procedure, the operation of the downstream reservoir was optimized using the historical inflow record to derive inflow transition probabilities. This step provided the initial release targets for the two upstream reservoirs. The operating strategies of each of the two upstream reservoirs were optimized separately for the previously derived release targets. Subsequently, the estimated releases from the upstream reservoirs were used as additional inflows to the downstream reservoir in a repeated optimization of its operation. For this purpose, a new set of transition probabilities was calculated, considering the changes in the inflow record. These iterative cycles were repeated until the stability of the overall system return was registered. The results obtained in the application of the methodology to the case study system showed close similarity to the observed historical system return. Slightly lower system benefits derived from the model are said to be the consequence of the limited precision of the SDP procedure, which was mainly due to the computational limitations on storage and inflow discretization.

Hall and Buras (1961) applied a three-level, DP based approach to solve a planning problem of capacity allocation among a number of reservoir sites. To reduce the dimensionality, they decomposed the original problem into three deterministic, hierarchically arranged, subproblems. At the first level, the objective was to identify a group of reservoir sites and their respective capacities by maximizing the overall system return. The second level optimization derived the optimal allocation of available releases among different uses for each of the selected reservoirs. Ultimately, using the former results, the water available for a particular use was optimally distributed among individual users. The solution to the overall problem was sought in a stepwise hierarchical manner starting from the first level. The results derived at a higher computational level were used as constraints at the immediate lower optimization level.

A major contribution to hierarchical multilevel decomposition approaches comes from Haimes (1977, 1982). The methodology is based on the decomposition of a complex system into smaller subsystems categorized into different levels of hierarchy. The principal idea behind the approach is to allow separate modeling and analysis at different decomposition levels. The information obtained at a certain decomposition level can then be further transmitted and used while analyzing the subsystems at the higher level of hierarchy. In general, hierarchical multilevel decomposition allows conceptual simplification of complex system modeling, which can result in a reduction of dimensionality. In addition, the analyst generally has to develop simpler computational and programming procedures and sometimes can even use the existing models.

5.1.2 Approaches based on aggregation/disaggregation principles

To solve high-dimensional optimization problems, aggregation/disaggregation aims to develop auxiliary models which are reduced in their complexity and which, at the same time, provide good approximations of the original problem. In most of the reported applications, a multiple-reservoir system was aggregated into a hypothetical single reservoir and the subsequent optimization was carried out for this simplified composite representation of the system. It is also quite common for aggregation/disaggregation methods to be used in combination with some decomposition principles to alleviate computational difficulties in optimization of complex reservoir system operation (Turgeon, 1980, 1981, and Archibald et al., 1997 in Section 5.1.1).

Rogers et al. (1991) presented the general concepts of aggregation/disaggregation methods and reviewed the respective applications in operations research. The authors emphasized several reasons in favor of using an aggregation/disaggregation modeling approach:

(a) It provides quick insight into the overall system structure and performance.

(b) A possible lack of reliable microlevel data may prohibit the development of a detailed model but, if the corresponding macrolevel data are available, it can also motivate the formulation of an aggregate model to analyze the problem on a larger scale.

(c) It enables analysts to obtain results at different levels of detail.

(d) The inherent computational burden can be significantly reduced.

Regardless of the mathematical programming and modeling techniques used in a particular application, a general formulation of an aggregation/disaggregation methodology comprises four principal steps. The first step involves the identification of pertinent data for aggregation and the subsequent process of combining them. This is followed by the creation of a composite model, which provides the reduction in complexity relative to the original model. Subsequently the analysis is carried out on the composite model and, at the final stage, the results derived for the hypothetical composite model are disaggregated into the respective components of the original problem. Although aggregation/disaggregation methods prove to be powerful tools for the dimensionality reduction of large-scale problems, they do require particular effort to be put into careful selection of the principles to be employed at each of the modeling stages in order to minimize the error induced by the simplification of the problem representation.

Saad and Turgeon (1988) proposed the principal component analysis (PCA) technique to reduce the number of state variables in the analysis of long-term multiple-reservoir operation problems. The PCA method was said to be applicable to problems where strong correlation between inflows to two or more reservoirs (or between reservoir storage states) could be detected. The procedure started by generating a set of synthetic streamflow sequences. Implicit SDP optimization followed to derive optimal operating strategies upon each generated inflow record. Subsequently, the PCA method was used to analyze the resulting policies and the achieved state variable values to find out whether the problem could have been modeled with fewer state variables. If so, the optimal operation policy for the reduced problem was derived by explicit SDP. The authors tested the applicability of the algorithm on a five-reservoir hydropower system on the La Grande River in Canada. In this particular case, the authors managed to reduce the original stochastic optimization problem of ten state variables to a four-state variable problem which was then solvable by DP.

Further improvements of the PCA method were reported in Saad *et al.* (1992), where the authors used censored-data statistical analysis to identify the parameters needed for PCA. The censored-data method provides the means to analyze a sample of observations for which it is known that the existing lower and/or upper bounds, if recorded a substantial number of times, can result in a biased estimate of the sample's probability distribution. With regard to the PCA method applied to a reservoir operation problem, these lower and upper bounds are the minimum and maximum storage volumes in the reservoirs observed from the sequence of deterministic optimizations performed over the set of synthetic streamflow sequences.

Using the same hydropower system as a case study, Saad *et al.* (1994) proposed a disaggregation approach based on the theory of neural networks. Initially, a five-reservoir system was aggregated into a single hypothetical reservoir whose operation was optimized by means of SDP. The composite operating strategy was subsequently disaggregated into individual reservoir policies using a feed-forward back-propagation neural network. The training of the neural network had been previously carried out over a large set of equally probable operating scenarios. To provide the training set, the authors generated a series of synthetic flow scenarios, which were further used to optimize the operation of the system assuming deterministic flow conditions. The application of the approach to the La Grande River hydropower system proved more efficient than the PCA method reported by Saad and Turgeon (1988).

Kularathna (1992) used aggregation/disaggregation methodology coupled with SDP based optimization to derive the operating strategy for the Mahaweli water resources system in Sri Lanka and that work is described in Section 5.2.7 in detail. The case study system consisted of three subsystems having three interlinked reservoirs each. First, each subsystem was represented by a single hypothetical composite reservoir. In the subsequent step the optimal operation policies were derived for the simplified system of three hypothetical reservoirs. Ultimately, the resulting operation policies of the composite reservoirs were decomposed into control rules of their respective individual reservoirs. The author found that the devised SDP based aggregation/disaggregation optimization approach produced a set of reservoir operating strategies which resulted in system performance very close to the deterministic optimum obtained by IDP. This work also included the application of two different decomposition techniques, i.e., the sequential and iterative decomposition algorithms, in optimization of multiple-reservoir systems operation. In addition, a comparison of explicit and implicit SDP optimization approaches was carried out on a reduced, three-reservoir subsystem. The conclusion drawn was that the explicit SDP model outperformed the implicit one. Such an outcome was put down to the inaccuracies incurred by the adopted streamflow generation model and the regression analysis.

5.1.3 Approaches based on continuous approximations of discrete functions

The basic idea behind this group of approaches is to tackle DP's curse of dimensionality by using a continuous rather than discrete representation of the objective function in order to allow a coarser discretization of the state space. This in turn enables the analyst to opt for a straightforward application of the chosen DP optimization approach, thus simultaneously considering all state variables of a multiple-state problem. Most of the reported studies have shown significant reductions in the number of discrete state values necessary to achieve acceptably low error levels of the objective function approximation. However, due to the fact that all state variables are considered simultaneously, the computational load imposed by these methodologies still increases exponentially with the number of state variables.

Murray and Yakowitz (1979) introduced constrained differential dynamic programming (CDDP) to operational analyses of multiunit reservoir systems under deterministic hydrological conditions. The proposed approach was actually a variation of IDP applied to all reservoirs of the system simultaneously. In order to avoid discretization of state and decision variables, the authors assumed that the objective function could be described by its continuous quadratic approximation. Therefore, the major task was to solve a quadratic programming problem at each stage, i.e., to minimize the quadratic function of multiple variables, subject to a set of imposed constraints. As a comparison to other approaches and to present the advantages of the method, CDDP was used to derive optimal policies for three characteristic multiple-reservoir system configurations: a four-reservoir system introduced by Larson (1968) and also used by Heidari et al. (1971) and Nopmongcol and Askew (1976); a four-reservoir system used by Chow and Cortes-Rivera (1974), and a hypothetical ten-reservoir system. Finally, as the main features of the algorithm, the authors emphasized fast convergence of CDDP, no need for discretization of state and decision space, and low computer storage, memory, and processing time requirements.

A similar idea was utilized by Foufoula-Georgiou and Kitanidis (1988) who introduced gradient dynamic programming (GDP) as a tool to solve optimal control problems of multiple-reservoir systems. In essence, GDP is backward moving DP carried out through temporal stages. The GDP approach allows simultaneous consideration of all system state variables by using the cubic hermite polynomial approximation of the objective function over state and decision space. The requirement of this approach is that the first derivatives of the interpolation polynomials must be continuous and known at each grid node. The method was tested on both deterministic and stochastic optimization problems for a four-reservoir system. In addition, the GDP algorithm was compared with the standard discrete DP procedure on a single-reservoir optimization problem. The results showed that the highly sophisticated mathematical procedure employed in GDP contributed to a significant reduction of the required state discretization level needed to achieve the acceptable accuracy of the results. In their earlier paper, Kitanidis and Foufoula-Georgiou (1987) compared the convergence rates of classical discrete DP and GDP and showed that, with the decrease of the state discretization interval, the GDP procedure converged more rapidly than conventional DP. The authors further expressed their belief that solutions to multiple-reservoir optimization problems should be sought in appropriate, case-dependent, interpolation-based numerical techniques rather than in discrete decomposition approaches. It is, however, arguable whether such an approach could be generally applicable since the computational load associated with GDP still increases exponentially with the number of state variables involved. Thus, and the obvious advantages GDP offers notwithstanding, the respective application of the method to very large reservoir systems would inevitably lead to the prohibitive increase of dimensionality, the well-known drawback of DP.

Johnson et al. (1993) proposed a high order piecewise polynomial approximation of the objective functions to allow a coarse discretization of the state space in multidimensional DP optimization problems. They used a piecewise cubic spline approximation of the objective function over the intervals created by state discretization. The coefficients of the cubic polynomials were derived upon the condition that they had to interpolate the objective function at each grid point of the state space. In addition, the first and second derivatives of the splines defined over the neighboring discretization intervals were required to be equal at the interval boundaries, thus providing the second-degree continuity of the approximation functions. The latter condition allowed the use of the quasi-Newton optimization algorithm to locate the extreme of the objective function approximation. The authors tested their approach on the same four-reservoir system used by Foufoula-Georgiou and Kitanidis (1988). They also carried out a comparison of the computation error and processing time requirements between their model, a piecewise linear approximation based model and GDP by Foufoula-Georgiou and Kitanidis (1988). A general conclusion was drawn that both the cubic spline model and GDP provided substantial processing time savings and error reduction as compared to the piecewise linear approximation model. The cubic spline based DP model was also found to achieve only a slightly smaller error for the same processing time than GDP. Tejada-Guibert *et al.* (1993) compared the same cubic spline DP and piecewise linear DP approaches on the two-reservoir Shasta/Trinity system in California and arrived at similar conclusions. Additional experiments showed that the proposed approach was successful in reducing the execution time for systems containing up to five reservoirs. However, a further increase in the number of reservoirs would have made the analysis susceptible to the curse of dimensionality for the number of state transitions still increased exponentially with the number of reservoirs in the system.

5.2 DECOMPOSITION METHOD

Decomposition approaches are very frequently used to alleviate dimensionality problems in operational analysis of large-scale systems. This chapter presents decomposition methods based on the principle of breaking down a multiple-reservoir system into individual reservoir units and a subsequent iterative determination of the individual reservoir operation policies. The three decomposition models developed are (a) sequential downstream-moving decomposition, (b) iterative downstream-moving decomposition, and (c) iterative up-and-downstream-moving decomposition. The models are applied to a multipurpose multiunit reservoir system in northern Tunisia (Milutin, 1998) and a multipurpose multiunit reservoir scheme in the Mahaweli river system in Sri Lanka (Kularathna, 1992).

5.2.1 Tunis water resources system

The Tunis water resources system analyzed in this study comprises seven reservoirs and a diversion weir located in the northern part of Tunisia as shown in Figure 5.1. However, note that the figure shows all reservoirs, including those planned and also those outside the system analyzed (which are not linked to each other with an interbasin transfer pipeline). In addition to water supply as their primary purpose, all of the seven reservoirs serve for flood protection, and some of them also have hydropower generation facilities. However, this presentation concentrates only on the long-term operational aspects of water supply, thus taking no account of flood protection and energy generation purposes. As Figure 5.2 reveals, the reservoirs interact by means of both serial and parallel interconnections. The available release from a reservoir may be distributed both to the local water users within its own basin, as well as towards remote users situated in other basins. The complexity of feasible water allocation patterns is reflected in the fact that one reservoir may provide water for a number of demand centers while, at the same time, a single demand may be supplied by more than one reservoir. The envisaged reservoir/demand links, together with the active storage capacities of the seven reservoirs are given in Table 5.1.

Time series of monthly inflow volumes for the seven reservoirs cover a period of 44 years (i.e. 1946–89). The average annual inflow to the entire system is estimated at $963.834 \times 10^6 \, m^3/$ year and the total active storage of the seven reservoirs amounts to $1000.7 \times 10^6 \, m^3$. Great variability of inflows under the prevailing semi-arid climatic conditions tends to constrain the utilization of the available resources. The magnitude of river flow variability can be observed from Tables 5.2 and 5.3. The two tables compile the historical mean incremental inflow data on monthly and annual scales for individual reservoirs as well as the respective total flow

Table 5.1. *Reservoir capacities and the associated demand targets*

Reservoir	In operation since	Capacity ($10^6 \, m^3$)	Targeted demand centers
Joumine	1983	121.3	BI, IMA, BLI, TU, TO, NA, MO, SO, SF
Ben Metir	1954	44.2	BE, JE, MB, TU
Kasseb	1968	72.2	TU
Bou Heurtma	1976	102.5	IBH
Mellegue	1954	89.0	INE, IBH
Sidi Salem	1981	510.0	IAEA, TU, TO, NA, MO, SO, SF, IBV, IMSC
Siliana	1988	61.5	ISI, IAEA, TU, TO, NA, MO, SO, SF, IBV, IMSC

All demand centers abbreviations starting with "I" refer to irrigation demand

Table 5.2. *Reservoir mean monthly incremental inflows (period 1946–89) ($10^6 m^3$/month)*

Reservoir	Sep.	Oct.	Nov.	Dec.	Jan.	Feb.	Mar.	Apr.	May	Jun.	Jul.	Aug.
Joumine	0.967	4.799	12.295	24.033	30.715	28.052	20.357	8.145	2.852	0.543	0.110	0.093
Ben Metir	0.287	0.864	2.925	6.825	8.974	9.445	7.224	4.189	0.920	0.303	0.181	0.187
Kasseb	0.682	1.551	3.699	7.956	11.104	8.136	6.935	4.897	1.643	0.687	0.587	0.512
Bou Heurtma	0.792	2.023	5.955	14.513	18.830	19.477	15.344	9.734	2.977	0.995	0.680	0.694
Mellegue	24.965	34.384	12.574	10.395	9.351	9.490	12.170	16.396	15.302	13.975	5.262	11.594
Sidi Salem	10.587	21.818	25.553	53.054	88.203	76.697	68.194	45.242	19.533	7.563	4.567	8.176
Siliana	3.456	5.610	3.431	3.690	5.164	5.157	5.548	4.484	2.496	1.282	1.049	1.732
System	41.736	71.049	66.432	120.466	172.341	156.454	135.772	93.087	45.723	25.348	12.436	22.988

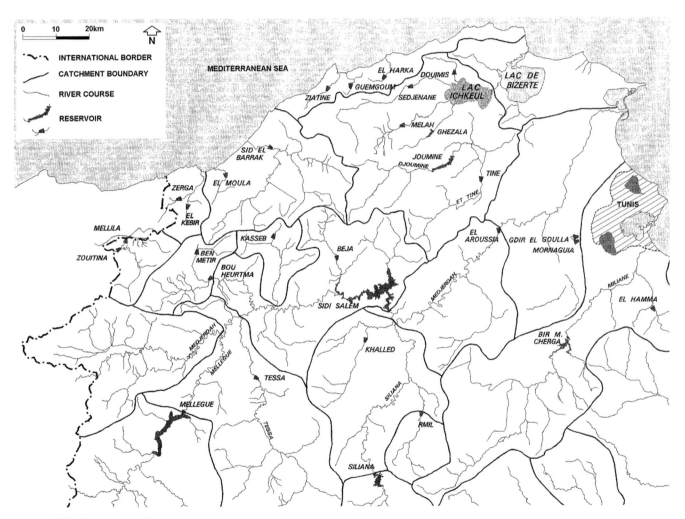

Figure 5.1 Tunis water supply system

availability for the entire system. Namely, the coefficient of variation of mean incremental annual inflows for the seven reservoirs varies between 0.481 for Kasseb and 0.968 for Siliana. On the system level, this statistical parameter is estimated at 0.465. As to the seasonal flow variability, the major portion of the system inflow (i.e., 84.6%) arrives in the period October–April, whereas the remaining 15.4% of the total is distributed over the period May–September. The three driest months on record are June, July, and August, jointly contributing only 6.3% of the total mean annual system inflow.

Table 5.3. *Basic statistics of the annual inflows for the seven reservoirs (period 1946–89)*

Reservoir	Range (10^6 m^3/year)	Mean (10^6 m^3/year)	σ (10^6 m^3/year)	Cv (–)
Joumine	20.100–300.720	132.959	79.494	0.598
Ben Metir	3.740–111.480	42.325	22.537	0.532
Kasseb	7.840–141.580	48.389	23.264	0.481
Bou Heurtma	9.360–245.320	92.015	48.494	0.527
Mellegue	53.640–804.850	175.859	125.335	0.713
Sidi Salem	152.778–1300.359	429.188	234.077	0.545
Siliana	7.520–201.960	43.099	41.715	0.968
System	335.261–2504.679	963.834	448.020	0.465

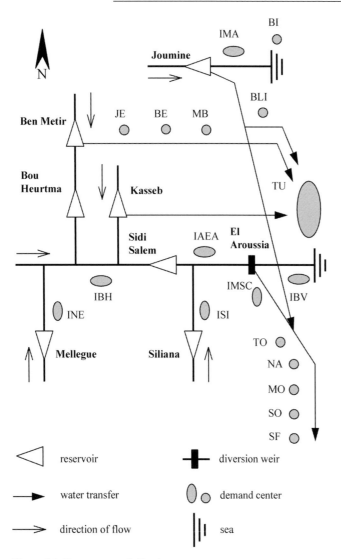

Figure 5.2 Seven-reservoir Tunis system

between 24 mm/month in January and 172 mm/month in July (Table 5.4).

The total estimated demand imposed upon the system is 469.504×10^6 m^3/year. This amounts to 48.7% of the mean annual inflow to the system. However, the unfavorable temporal and spatial distribution of demands and available inflows still poses a considerable obstacle for successful operation of the system. For instance (see Table 5.5), the driest three-month period from June to August is characterized by the total system demand of 210.155×10^6 m^3, which is 44.7% of the total annual demand. On the other hand, the mean available inflow volume in the same period reaches only 60.772×10^6 m^3, which amounts to 6.3% of the mean annual inflow into the system given in Table 5.2. These figures clearly reveal that Tunisia has no other options but storage and interbasin transfer to meet its water demands.

Water demand imposed upon the system is partitioned among 18 individual demand centers shown in Table 5.5. There are four distinct water uses considered: drinking water demands of various municipal areas; a specific drinking water demand of large tourist centers along the Mediterranean coast; irrigation demands; and a recharge of one natural lake (Lac Ichkeul) located in the north of the country. The major water users in the system are the drinking water demand TU, the tourist centers' water demand TO, the requirement for the natural lake recharge BLI and the IAEA, IBV, IMSC, IBH, and ISI irrigation demands. These eight demand centers constitute 94.1% of the total demand imposed upon the system.

As Table 5.1 reveals, a considerable number of the demand centers receive water from more than one reservoir. Namely, in four cases there are two reservoirs supplying a demand center (i.e., irrigation water demands IBH, IAEA, IBV, and IMSC), whereas five demand centers (i.e., drinking water demands NA, MO, SO, SF, and the tourist centers' water requirement TO) receive water from three reservoirs, and only one user (i.e., drinking water demand TU) gets water

In addition to inflow variability, the efficient exploitation of the available water is further limited by water losses, mainly due to evaporation. For instance, the mean monthly elevation losses due to evaporation estimated for the entire system vary

Table 5.4. *Estimated mean monthly elevation losses due to evaporation (mm/month)*

Reservoir	Sep.	Oct.	Nov.	Dec.	Jan.	Feb.	Mar.	Apr.	May	Jun.	Jul.	Aug.
Joumine	132	66	34	27	25	22	36	47	105	159	227	204
Ben Metir	111	72	44	33	31	31	53	64	99	130	162	152
Kasseb	111	55	29	22	21	18	30	40	89	135	191	172
Bou Heurtma	79	51	25	22	20	20	32	33	61	99	131	125
Mellegue	99	57	32	21	21	25	46	61	99	132	166	155
Sidi Salem	120	77	48	36	33	34	57	69	107	141	176	164
Siliana	88	59	34	25	20	22	33	42	75	114	148	129
Average	106	62	35	27	24	25	41	51	91	130	172	157

Table 5.5. *Monthly water demands for the 18 demand centers in northern Tunisia (10^6 m³/month)*

Demand	Sep.	Oct.	Nov.	Dec.	Jan.	Feb.	Mar.	Apr.	May	Jun.	Jul.	Aug.
TU	4.634	4.538	4.305	4.181	4.529	4.048	4.515	4.691	4.977	5.234	5.734	5.738
MO	0.101	0.090	0.082	0.079	0.069	0.065	0.084	0.094	0.100	0.110	0.129	0.128
NA	0.116	0.099	0.086	0.081	0.080	0.073	0.090	0.098	0.108	0.119	0.145	0.150
SO	0.290	0.275	0.252	0.229	0.205	0.190	0.250	0.264	0.274	0.296	0.355	0.372
SF	0.762	0.654	0.626	0.567	0.544	0.484	0.615	0.675	0.715	0.739	0.766	0.801
BI	0.245	0.233	0.217	0.215	0.231	0.197	0.220	0.239	0.250	0.250	0.278	0.289
JE	0.108	0.103	0.092	0.090	0.096	0.084	0.096	0.099	0.104	0.106	0.105	0.130
BE	0.136	0.125	0.120	0.119	0.105	0.108	0.128	0.128	0.137	0.147	0.166	0.162
MB	0.039	0.036	0.034	0.034	0.030	0.031	0.037	0.037	0.039	0.042	0.047	0.046
IMA	0.172	0.188	0.228	0.000	0.000	0.138	0.644	0.489	0.934	1.132	0.771	0.714
BLI	0.122	0.609	1.913	3.874	4.741	4.147	2.943	1.211	0.352	0.063	0.015	0.011
TO	1.624	1.459	0.914	0.777	0.716	0.765	1.138	1.211	1.388	1.531	1.892	2.077
IAEA	2.881	2.131	1.391	0.000	0.000	0.235	2.287	4.820	7.488	9.350	10.001	7.519
IBV	1.250	0.627	0.305	0.002	0.002	0.003	0.154	0.768	1.813	2.736	3.089	2.715
IMSC	12.784	5.420	2.258	0.606	0.505	2.020	6.586	7.675	13.132	21.709	26.502	25.250
INE	0.204	0.079	0.060	0.000	0.000	0.000	0.070	0.119	0.224	0.487	0.654	0.585
IBH	20.154	10.369	6.283	0.000	4.363	6.611	7.790	16.018	10.660	16.661	21.693	13.170
ISI	2.360	1.308	0.676	0.000	0.042	0.975	0.862	1.770	4.286	5.619	6.227	5.380
Total	47.982	28.343	19.842	10.854	16.258	20.174	28.509	40.406	46.981	66.340	78.578	65.237

from five reservoirs. These ten demands amount to 86.5% of the total annual demand. The remaining eight demand centers depict the reservoirs' local users (NB 13.5% of the total demand imposed upon the system): irrigation schemes IMA, INE, and ISI; drinking water demands BI, JE, BE, and MB; and water requirements for recharge of Lac Ichkeul (BLI).

The largest reservoir in the system, Sidi Salem, is situated on the Medjerdah River and represents the backbone of the seven-reservoir system. Its capacity amounts to 51.0% of the total system active storage while, on an average annual scale, its incremental inflow reaches 44.5% of the total inflow to the system. Sidi Salem also regulates and utilizes any excess release that may originate from the three reservoirs situated directly upstream: Kasseb, Bou Heurtma, and Mellegue. The users associated with this reservoir include almost all of the major demand centers in the system. Some of the associated demand centers are located immediately downstream of the reservoir (i.e., IAEA, TU, and IBV), whereas the rest of them (i.e., TO, NA, MO, SO, SF, and IMSC) get water via the Medjerdah–Cap Bon Canal, which departs from the Medjerdah River at diversion weir El Aroussia downstream of Sidi Salem, as shown in Figure 5.2.

Siliana is a small reservoir located on the river Siliana in the Medjerdah basin. Although the inflow to this reservoir is poor (i.e., 4.5% of the total annual inflow into the system), the list of potential Siliana users is long: ISI, IAEA, TU, TO, NA,

MO, SO, SF, IBV, and IMSC. Nevertheless, the main purpose of Siliana is to provide water for the local irrigation scheme ISI. To a certain extent, Siliana is expected to compensate for any potential supply shortage that may occur in the operation of other reservoirs supplying the common demands on the list of Siliana's users. Thus, part of the Siliana release may also be conveyed through the Medjerdah–Cap Bon Canal.

Unlike Siliana, the Joumine Reservoir can contribute significantly to the supply of all the associated demand centers. It is located on the Joumine River in the far north of the country. This is the only reservoir in the system which is not located in the immediate Medjerdah River basin. Its mean annual inflow amounts to 13.8% of the total system water resources. In addition to the three local users (BI, IMA, and BLI), Joumine plays an important role in providing water for the remaining remote demand centers (TU, TO, NA, MO, SO, and SF) which it supplies jointly with other reservoirs from the system. The remote users get water allocated from Joumine via a pipeline, which at its end discharges into the Medjerdah–Cap Bon Canal.

The Bou Heurtma Reservoir serves primarily for irrigation water supply of its local demand IBH. It is located on the Bou Heurtma River, a tributary of the Medjerdah, with a mean unregulated inflow to the reservoir of 9.5% of the total annual inflow into the system. Furthermore, Bou Heurtma can accommodate and regulate any excess release that may be produced by its upstream counterpart Ben Metir. In addition to its consumptive demand IBH, Bou Heurtma may contribute to regulate the inflow to Sidi Salem.

The Mellegue Reservoir is located on the Mellegue tributary of the Medjerdah River. The mean unregulated inflow to Mellegue amounts to 18.2% of the total system inflow. However, this is the reservoir with by far the lowest reservoir volume factor among all the reservoirs in the system. Its capacity hardly reaches 50.6% of the respective mean annual inflow. As a comparison, the next highest volume factor is that of Joumine (0.912), whereas the highest one is related to Kasseb: 1.492. As to the associated water users, Mellegue provides water for the local irrigation scheme INE and, jointly with Bou Heurtma, covers the IBH irrigation demand. Like Bou Heurtma, Mellegue can also contribute to the increase of Sidi Salem's inflow during the peak demand periods.

The Kasseb Reservoir is situated on the Kasseb River in the Medjerdah basin. Its mean annual incremental inflow is at the level of only 5.0% of the total system inflow. The sole purpose of this reservoir is to provide drinking water for the TU demand center. Any excess release from Kasseb may be used to cover the supply shortage of Sidi Salem.

Ultimately, Ben Metir is the smallest reservoir in the system with equally small incremental inflows (i.e., only 4.4% of the

Table 5.6. *Capacities of the existing water conveyance structures*

Water conveyance structure	Capacity	
	(m³/s)	(10⁶ m³/month)[a]
Medjerdah–Cap Bon Canal (MCB)	16.0	42.163
Joumine–MCB pipeline	4.0	10.541
Kasseb–Tunis pipeline	1.1	2.899
Ben Metir–Tunis pipeline	1.0	2.635

[a] Estimated assuming the average number of days in a month to be 30.5

total system inflow). It is located on the El Lil River in the immediate basin of the Bou Heurtma River. Ben Metir contributes primarily towards drinking water supply of its local users JE, BE, and MB and provides water for the TU demand. Since it is located upstream of Bou Heurtma, if any surplus water is available, Ben Metir is expected to compensate for the potential shortage of Bou Heurtma's water deliveries.

There are a number of water conveyance (transfer) structures in the system. The capacities of major conveyance structures are given in Table 5.6.

5.2.2 System decomposition: Tunis system

The three decomposition methods described in the subsequent three sections are based on the principle of breaking down a multiple-reservoir system into individual reservoir units and a subsequent iterative determination of the individual reservoir operation policies. To derive the operation policies of individual reservoirs, each of the methods employs the same iterative six-step optimization/simulation procedure, which involves:

(1) estimation of the inflow into a reservoir;
(2) evaluation of the demand imposed upon a reservoir;
(3) stochastic dynamic programming based optimization of the operation of a reservoir;
(4) simulation of the reservoir's operation according to the derived SDP policy;
(5) the total releases obtained by simulation are allocated to individual users; and
(6) estimation of the expected unmet demands and the expected total supply deficits associated with the operation of the reservoir in question.

With regard to the relative flexibility of the basic decomposition principles, there generally exist a number of possible reservoir orderings which comply with the imposed decomposition rules. However, the operational analysis of the Tunis system has been limited to only three alternative reservoir sequences, each

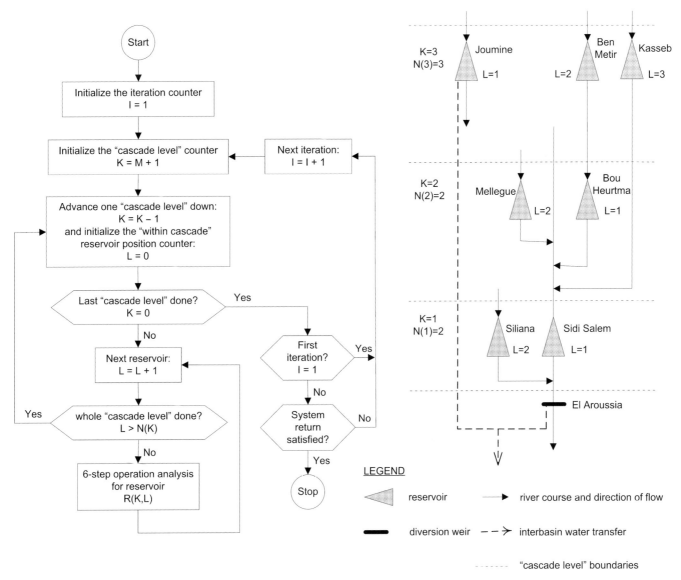

Figure 5.3 Sequential downstream-moving decomposition flow chart and Tunis system

exemplifying a distinct decomposition approach. What distinguishes these three decomposition approaches from one another is the way they address the problem of modeling the interaction among serially linked reservoirs. Sections 5.2.3 to 5.2.5 introduce the three employed decomposition approaches and present their respective applications to the Tunis system.

5.2.3 Sequential downstream-moving decomposition: Tunis system

The ordering of individual reservoirs generally follows the direction of river flows in the river basin(s). Within each iteration, the analysis starts from the uppermost reservoir in the system. Thereafter, the selection of reservoirs proceeds in

the downstream direction until all the reservoirs have been taken into consideration, which completes one iterative cycle. Such cycles are then repeated until a satisfactory stabilization of the total system return has been achieved.

Figure 5.3 presents the general flow chart of the applied sequential downstream-moving decomposition (SDD) and the symbolic representation of the decomposed Tunis system according to this approach. Both figures utilize the same notation, which is described with the introduction of the main SDD decomposition features. The definition of reservoir ordering is based on two principles:

(a) Reservoirs are initially clustered into cascade levels (K) to distinguish between subsets of reservoirs with respect to the sequence upon which those subsets will be entering the principal iterative cycles (*I*). According to the SDD approach, the cascade level ordering is guided by the descending arrangement of their respective indices *K*. The total number of cascade levels is represented by the parameter *M* in Figure 5.3.

(b) Reservoir selection order (*L*) within a cascade level can be defined by any rules imposed by the analyst. These may include firm water allocation schemes, water quality requirements, or some empirical rules based, for instance, on operating or environmental issues.

The reservoir ordering for the case study system in the SDD decomposition approach is: Joumine, Ben Metir, Kasseb, Bou Heurtma, Mellegue, Sidi Salem, and Siliana. The choice of the Joumine–Ben Metir–Kasseb ordering in cascade *K* = 3 is generally an arbitrary one. On the other hand, the Bou Heurtma–Mellegue ordering in cascade *K* = 2 is determined by the priority Bou Heurtma has towards supplying their joint irrigation demand IBH. The Sidi Salem–Siliana sequence in cascade *K* = 1 is solely based on the superior size, water availability, and principal role Sidi Salem exhibits in the system.

The adopted decomposition methodology relies on the iterative analyses of individual reservoir operations to arrive at the operating strategy of the entire system. The approach maintains the interactions among reservoirs within its iterative process. The SDD decomposition utilizes three principles to this end. Namely, upon completing the analysis of the operation of a reservoir, three distinctive pieces of information are made available for further analyses:

(a) Within one iterative cycle, the estimated expectations of monthly demands, which have not been covered so far, are regularly updated after each reservoir's operating strategy has been derived by optimization and appraised by simulation. That is, the operation of the next reservoir in the sequence is analyzed with respect to the updated expectation of the system's demand records. For instance, being the first in the optimization sequence, Joumine faces the entire TO demand. Upon estimating the expected monthly allocation of Joumine resources to this demand, any of the expected unmet monthly TO requirements are to be associated with Sidi Salem, the next reservoir in sequence, to supply this demand. Ultimately, the operation of Siliana is optimized taking into account the expected remaining part of the TO demand which could not have been covered by Joumine and Sidi Salem.

(b) The aggregate of the expected monthly estimates of all the unmet demands associated with a particular reservoir is regarded as the total expected supply deficit of that reservoir. In the subsequent iteration cycle, the monthly estimates of a reservoir's supply deficits are used as an additional, hypothetical demand imposed upon the reservoirs situated directly upstream of the reservoir in question. Consequently, the upstream reservoirs' operating strategies derived in the succeeding iteration would be altered to try to release additional water to increase the inflow into the downstream reservoir in those periods when the operation of the downstream reservoir exhibits supply shortage. With regard to the Tunis system, for example, the total consumptive demand imposed upon Ben Metir is increased by the expected supply deficit of Bou Heurtma estimated in the preceding iteration. Similarly, the previous iteration supply deficit of Sidi Salem is associated with Kasseb, Bou Heurtma, and Mellegue (being covered by more than one reservoir, the expected supply deficit of Sidi Salem is also subject to demand updating as described in the previous point).

(c) Upon allocating water to all the associated users, the remaining part of the total reservoir release, if any, is considered as a supplementary inflow to the reservoir located immediately downstream. For instance, Bou Heurtma's incremental inflows are increased by the non-utilized releases from Ben Metir estimated in the same iterative cycle and, similarly, Sidi Salem makes use of the additional inflow originated from excess releases from Kasseb, Bou Heurtma, and Mellegue, all obtained in the same iteration.

5.2.4 Iterative downstream-moving decomposition: Tunis system

Iterative downstream-moving decomposition (IDD) is essentially a variation of the SDD approach. Consequently, reservoir ordering in IDD is also determined based on the two principles related to the cascade level definitions and within-cascade reservoir sequences. In addition, the interactions among the serially connected reservoirs and demand updating are defined in the same way as within the SDD approach. The flow chart of the IDD decomposition and the Tunis system are given in Figure 5.4. The notation used to describe the IDD decomposition is identical to the one introduced in Section 5.2.3, with the only addition being the cascade attribute *D* whose role is clarified in the following passage.

The formulation of the IDD approach brings about an improvement to the way the SDD decomposition deals with the cases where several reservoirs in parallel are serially linked to one reservoir situated downstream of them. In SDD the analyses of the operation of all the reservoirs on one cascade level are completed before proceeding to the next downstream

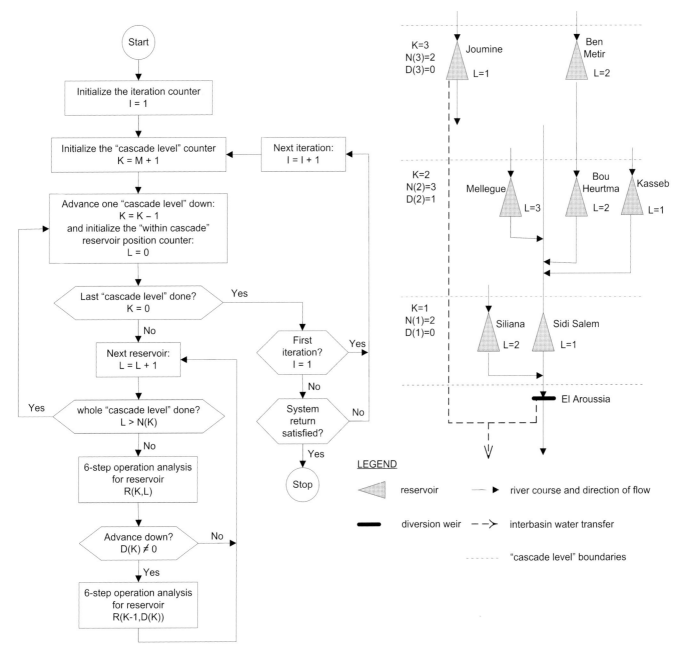

Figure 5.4 Iterative downstream-moving decomposition flow chart and Tunis system

cascade level. Thus, if the upper cascade level contains a number of reservoirs which can contribute to the increase of the inflow to one of the reservoirs on the lower cascade level, the optimization of the operation of the downstream reservoir is carried out only after the analyses of all of its direct upstream counterparts have been completed. Since the interactions among serially connected reservoirs are approximated by the exchange of information about the expected supply deficits of the downstream reservoir obtained in the preceding

iteration and the time series of nonutilized flows from the upstream reservoirs derived in the present iterative cycle, it is obvious that the estimation of the expected supply deficit of the downstream reservoir can repeatedly be updated after completing the analysis of the operation of each of the upstream reservoirs. In other words, the time series of the excess flows from the first analyzed upstream reservoir can be used to make the initial update of the downstream reservoir's operation policy and, in turn, to re-evaluate its expected supply deficits. Thereafter, the updated expected supply deficits are used as additional hypothetical demand imposed upon the next upstream reservoir. This process is repeated for each of the upstream reservoirs which are serially connected to the downstream one.

According to the IDD decomposition, the sequence upon which the reservoirs enter the computational process within one iterative cycle is: Joumine, Ben Metir, Kasseb, Sidi Salem, Bou Heurtma, Sidi Salem, Mellegue, Sidi Salem, Siliana. The operation of Sidi Salem is derived in three consecutive steps following the optimization of Kasseb, Bou Heurtma, and Mellegue. This process is controlled by the introduction of the cascade level attribute D identifying the reservoir from the immediate downstream cascade level whose operation is to be repeatedly optimized following the analysis of each of the reservoirs from the present cascade level (e.g., the value of D for cascade level $K = 2$ is $D(2) = 1$, which points to Sidi Salem whose index in the immediate downstream cascade is $L = 1$).

5.2.5 Iterative up-and-downstream-moving decomposition: Tunis system

Iterative up-and-downstream-moving decomposition (UDD) departs from the former two decomposition methods in the sense that the adopted reservoir sequence generally follows the direction opposite to the direction of river flows. On the other hand, the common feature among the three is the principle of breaking down a complex system into individual reservoirs by identifying reservoir subsets at different cascade levels with the subsequent determination of within-cascade reservoir analysis orders. However, unlike the SDD and IDD methods, the UDD decomposition analyzes the individual reservoir operations by starting from the lowest cascade level and thereafter proceeding upstream along the cascade levels. In addition, any of the existing serial reservoir links are modeled by an iterative up-and-down progression within the respective subset of reservoirs. The flow chart of the UDD decomposition is depicted in Figure 5.5. The case study system (Tunis system) decomposition according to this method is presented in Figure 5.6. The general cascade level and reservoir position notation used in the UDD decomposition is identical to the one given for the former two methods. Some additional system decomposition attributes, i.e., the parameter U identifying the number of upstream reservoirs serially linked to a particular reservoir and the vector of indices V of the respective upstream reservoirs, are described below.

The individual reservoir operation analysis within one iteration of the UDD decomposition starts from the lowest cascade level in the system. The information interchange between two subsequent iterations is, unlike in SDD and IDD

approaches, the set of time series of nonutilized flows from the reservoirs. These records are used as additional inflows into the respective downstream reservoirs in serially linked reservoir clusters, if any. If the reservoir whose operating analysis has just been completed is serially linked to any number of reservoirs from the upstream cascade level (i.e., the attribute U for the reservoir is not zero), the process continues by advancing to the upstream cascade level to analyze the operation of those reservoirs, V, which are linked to the reservoir in question. The operations of those reservoirs are then optimized and simulated taking into account the expected monthly supply deficits of their downstream counterparts. Upon completing the upstream cascade analyses, the process returns to the downstream reservoir where it has made an advance in the upstream direction. At this point, the optimization and simulation of the operation of this reservoir are carried out once again. This is done to update its operating strategy by taking into account the additional inflow time series obtained in the analyses of the reservoirs from the upstream cascade level. Once such an iterative up-and-down analysis is completed for a serially linked cluster, the process continues with the next reservoir in the presently lowest cascade whose analysis has not been completed yet. Similarly to the other two decomposition methods, the UDD decomposition also observes the demand updating principle in addition to the exchange of information about the nonutilized releases and the expected monthly supply deficits.

Perhaps the best way to clarify this description is to apply the UDD principles to the case study system (Figure 5.6). Thus, the reservoir sequence in an iteration of the UDD decomposition is: Joumine, Sidi Salem, Kasseb, Bou Heurtma, Ben Metir, Bou Heurtma, Mellegue, Sidi Salem, and Siliana. It can be clearly seen that the iterative process of the analysis of serially linked reservoirs is recursive. Namely, the outer cluster with the Sidi Salem Reservoir as the downstream one contains two serially linked reservoirs: Bou Heurtma and Ben Metir. Therefore, upon reaching Bou Heurtma in the process of analyzing Sidi Salem's upstream counterparts, the analysis is held up until the Bou Heurtma–Ben Metir serial link is completed. This is indicated by different shading patterns used to identify the respective reservoir clusters (Figure 5.6) and by the flow chart of the recursive part of the algorithm given in Figure 5.5.

5.2.6 Comparison of the three decomposition alternatives: Tunis system

This section presents the outcomes of the optimization and the respective simulation analyses of the long-term operation of the Tunis system executed for the three alternative

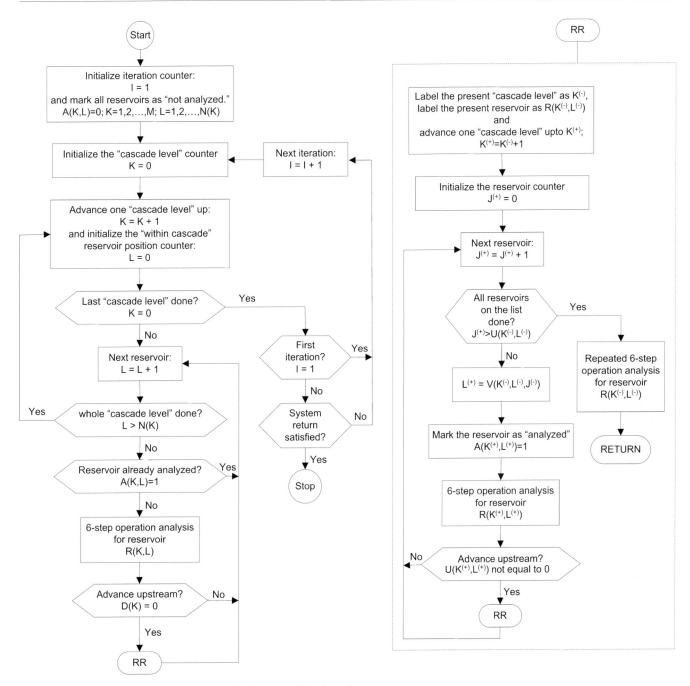

Figure 5.5 Iterative up-and-downstream-moving decomposition flow chart

decomposition approaches. The three **SDP** based decomposition models share a number of common features.

The number of characteristic discrete storage representations is set to 25 for each reservoir in the system. Table 5.7 displays the adopted discrete storage representations for individual reservoirs. It should be noted here that the sequential downstream-moving decomposition (SDD) was also tested using 48 storage

classes (i.e., achieving a 50% reduction of the respective class sizes obtained with 25 discrete storage representations) producing almost no improvement of the system's operation as compared to the adopted coarser discretization level.

Monthly reservoir inflows are represented by the respective sets of discrete flow values. The maximum allowed number of discrete inflow classes is set to 12. Inflow discretization varies

Table 5.7. *Discrete storage representation for individual reservoirs ($10^6 m^3$)*

Class	Joumine	Ben Metir	Kasseb	Bou Heurtma	Mellegue	Sidi Salem	Siliana
1	130.0	57.2	81.9	117.5	120.0	555.0	70.0
2	127.4	56.2	80.3	115.3	118.1	543.9	68.7
3	122.1	54.3	77.2	110.8	114.2	521.7	66.0
4	116.8	52.4	74.1	106.4	110.3	499.6	63.3
. . .	. . .	. . .	. . .	. . .	. . .	. . .	. . .
22	21.9	17.8	17.5	26.1	40.7	100.4	15.2
23	16.6	15.9	14.4	21.7	36.8	78.3	12.5
24	11.3	14.0	11.3	17.2	32.9	56.1	9.8
25	8.7	13.0	9.7	15.0	31.0	45.0	8.5

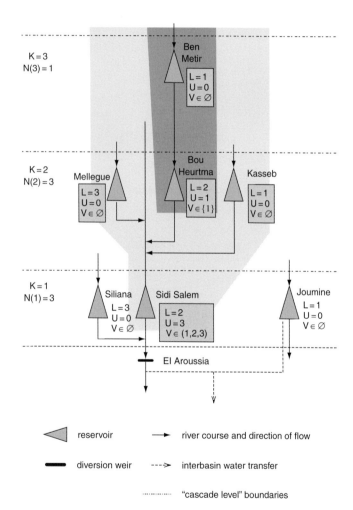

Figure 5.6 Iterative up-and-downstream-moving decomposition of the Tunis system

the inflow classes for each month, the stochasticity of monthly inflows into reservoirs is described by transition probabilities estimated for their respective discrete representations based on the 33-year-long policy determination inflow subsets.

The objective function used in optimization of the operation of individual reservoirs is to minimize the expectation of the annual aggregate of the squared deviation of a monthly release from the respective target.

The adopted simulation alternative within each of the three decomposition approaches assumed full compliance with the derived SDP policies, thus allowing no policy violations. The individual reservoir operation policies derived by SDP within each of the three decomposition approaches are defined for each month within an annual cycle. They are given in the form of a table indicating the class index of the recommended final storage volume as a function of the class indices of the initial storage volume and inflow for that particular month. An abridged example of a typical SDP policy table is presented in Table 5.8 (note that storage volume decreases with the increase of the storage class index whereas inflow volume increases with the increase of the inflow class index).

Table 5.9 presents the number of iterations required to achieve stabilization of the termination criterion (i.e., the expected annual supply deficit of the entire system). The UDD is the fastest among the three models. According to the estimates of the expected annual supply deficit of the entire system, the policies derived by the three decomposition models result in virtually identical system performances. The obtained values show that the expected annual water supplies vary between 95.8% (SDD and UDD) and 95.9% (IDD) of the annual demand.

The estimated expected annual deficits of individual demands show that SDD and IDD models outperform UDD by a narrow margin as presented in Tables 5.9 and 5.10.

from reservoir to reservoir and from month to month. The number of discrete inflow representations is defined as a linear function of the reservoir's capacity and the range of inflow observations in that particular month. Upon defining

Table 5.8. *An example of a typical SDP based operation policy table*

Initial storage class	Inflow class				
	1	2	3	4	5
1	6	6	1	1	1
2	7	6	2	1	1
3	8	7	3	1	1
4	9	8	4	2	1
...	...	...	...	...	...
22	24	23	19	18	15
23	24	23	20	19	16
24	25	24	21	20	17
25	25	25	22	20	17

Table 5.9. *Comparison of the three decomposition alternatives*

Decomposition method	Number of iterations	Expected annual supply deficit of entire system (10^6 m^3)
SDD	6	19.736
IDD	3	19.117
UDD	2	19.786

Table 5.10. *Expected annual deficits of individual demand centers for SDD, IDD, and UDD models (10^6 m^3/year)*

Demand center	Expected annual deficit (10^6 m^3)		
	SDD	IDD	UDD
TU	0.293	0.294	0.288
MO	0.022	0.022	0.023
NA	0.024	0.024	0.025
SO	0.065	0.065	0.066
SF	0.163	0.163	0.182
BI	0.065	0.065	0.065
JE	0.054	0.054	0.050
BE	0.052	0.051	0.050
MB	0.021	0.023	0.019
IMA	0.137	0.137	0.137
BLI	1.088	1.088	1.088
TO	0.282	0.282	0.283
IAEA	0.142	0.142	0.116
IBV	0.100	0.090	0.134
IMSC	5.524	5.413	5.810
INE	0.007	0.003	0.003
IBH	5.681	5.536	5.785
ISI	5.656	5.664	5.662

The simulation shows that the system is likely to fail more severely during dry summer months (i.e., June–August) when the imposed demand for water is at its peak. However, the derived SDP policies manage to reduce the magnitude of supply deficits in these months by "spreading" the inevitable shortage over the remaining nine months of an annual cycle. The largest "share" of this shortage is associated with the immediate neighboring months (i.e., April–May and September–October).

Table 5.11 shows, however, that the three decomposition models achieve similar performances of the system on the basis of different operation policies. The only reservoir whose SDP policies do not differ from one model to another is Joumine. This is because, in all of the models, Joumine is the first reservoir to be considered in the computational process, thus always having the same input sets of hydrologic and demand variables irrespective of the chosen decomposition model.

5.2.7 Sequential downstream-moving decomposition: Mahaweli system

This section presents the application of SDD to the Mahaweli water resources system in Sri Lanka. The system is described in detail in Section 2.2.1.

Sequential optimization is initiated with the optimization of the uppermost reservoir subsystem, which consists of Caledonia, Talawakelle, and Kotmale Reservoirs. For this optimization, the SDP model described in Section 3.2 is used with necessary modifications. The power plant of the Talawakelle Reservoir is considered as a run-of-the-river power plant due to the small storage capacity of the Talawakelle Reservoir. Simulation of the operation of the same subsystem using available historical monthly streamflow records is carried out according to the operation policies derived by optimization and the optimal diversion policy of the Polgolla Barrage as described in Section 5.3 (q.v.). It results in an operation pattern of the reservoir subsystem, monthly diversions and releases at Polgolla. These diversions and releases become the upstream inflows to the Ukuwela–Bowatenna–Moragahakanda and Victoria–Randenigala–Rantembe reservoir subsystems respectively.

The operation of the two downstream reservoir subsystems is then optimized individually, considering inflows contributed by the Polgolla Barrage in addition to the inflows within the respective subsystems. In the optimization of the Victoria–Randenigala–Rantembe subsystem, the Rantembe power plant is considered as a run-of-the-river power plant due to the small storage capacity of the Rantembe Reservoir. After formulating operation policies by optimization, the operations of two downstream subsystems are then simulated

Table 5.11. *SDD, IDD, and UDD models: relative number of different decisions in monthly policy tables (%)*

Reservoir	SDD vs. IDD			SDD vs. UDD			IDD vs. UDD		
	min	mean	max	min	mean	max	min	mean	max
Joumine	0.0	0.0	0.0	0.0	0.0	0.0	0.0	0.0	0.0
Ben Metir	1.0	5.0	14.0	1.3	11.6	40.0	4.0	14.4	39.3
Kasseb	0.0	1.8	7.2	0.7	6.3	16.0	0.7	5.9	14.7
Bou Heurtma	1.1	8.6	21.0	6.0[a]	22.4[a]	40.0[a]	7.0[a]	20.5[a]	51.3[a]
Mellegue	1.3	3.4	6.3	0.8	4.8	13.1	0.0	5.1	14.7
Sidi Salem	0.0	4.8	20.7	1.6	14.7	40.7	3.2	14.5	46.7
Siliana	0.0	0.6	4.0	0.0	3.0	10.7	0.0	2.7	6.7

[a] Some months are excluded from the comparison because the respective policies have different numbers of inflow classes

independently using available historical flow records and the derived operation policies.

The mathematical formulation of the sequential optimization approach (consisting of three SDP based optimization models) is presented in detail for the Mahaweli system. The objective function is to maximize the expected energy generation. Using the usual notation (Section 3.2) and the superscripts $N = 1, 2,$ and 3 to represent the CTK, UBM, and VRR subsystems respectively (Figure 2.11), and the subscripts $i = 1, 2, 3$ to represent the three reservoirs/power plants (starting from the upstream) in each subsystem:

For the CTK subsystem,

$$\text{Maximize } \xi \left\{ \sum_{j=1}^{T} \left(\sum_{i=1}^{3} \text{TEP}_{i,j} \right) \right\}, \tag{5.1}$$

where

$$\text{TEP}_{i,j} = 9.81 \times \eta \times R_{i,j}^1 \times (\text{EL}_{i,j}^1 - \text{DWL}_{i,j}^1) \times t_j / 10^6 (\text{MWh});$$
$$i = 1, 2, 3; \qquad j = 1, 2, \dots, 12. \tag{5.2}$$

State transformation equation for the Caledonia reservoir:

$$S_{1,j+1}^1 = S_{1,j}^1 + I_{1,j}^1 - E_{1,j}^1 - R_{1,j}^1 - O_{1,j}^1; \qquad j = 1, 2, \dots, 12. \tag{5.3}$$

State transformation equation for the assumed run-of-the-river power plant at Talawakelle:

$$R_{2,j}^1 = R_{1,j}^1 + O_{1,j}^1 + I_{2,j}^1; \qquad j = 1, 2, \dots, 12. \tag{5.4}$$

State transformation equation for the Kotmale reservoir:

$$S_{3,j+1}^1 = S_{3,j}^1 + I_{3,j}^1 - E_{3,j}^1 - R_{3,j}^1 + R_{2,j}^1 - O_{3,j}^1;$$
$$j = 1, 2, \dots, 12, \tag{5.5}$$

$$Q_{p,j} = R_{3,j}^1 + O_{3,j}^1 + \text{IP}_j; \qquad j = 1, 2, \dots, 12. \tag{5.6}$$

Where

$Q_{p,j}$ = inflow volume at interface point (Polgolla Barrage) during period j ($10^6 \, \text{m}^3$),

IP_j = incremental inflow to Polgolla Barrage during period j ($10^6 \, \text{m}^3$).

In addition to the constraints imposed by the release and storage limits of the reservoirs, optimization is subject to the following constraint:

$$Q_{p,j} \geq Q_{p,j}^*, \tag{5.7}$$

$Q_{p,j}^*$ = optimum inflow at Polgolla obtained by three-composite-reservoir optimization model ($10^6 \, \text{m}^3$) (in Section 5.3.2).

For the SDP models of the downstream subsystems, the same form of the objective function is used. The downstream irrigation water demands are considered as constraints. The upstream (simulated) inflows are defined as in the following.

For the UBM system:

$$Q_{u,j} = D(Q_{p,j}^s); \qquad j = 1, 2, \dots, 12, \tag{5.8}$$

where

$Q_{p,j}^s$ = inflow at Polgolla in period j obtained by simulating (superscript s indicates this) CTK subsystem according to its optimum operation policies ($10^6 \, \text{m}^3$),

$D(.)$ = representation of diversion policy at Polgolla, and

$Q_{u,j}$ = inflow entering UBM subsystem across interface point at Polgolla in period j (determined according to the Polgolla diversion policy) ($10^6 \, \text{m}^3$).

For Ukuwela power plant:

$$R_{1,j}^2 + O_{1,j}^2 = Q_{u,j}; \qquad j = 1, 2, \dots, 12. \tag{5.9}$$

For Bowatenna Reservoir:

$$S_{2,j+1}^2 = S_{2,j}^2 + R_{1,j}^2 + O_{1,j}^2 + I_{2,j}^2 - E_{2,j}^2 - R_{2,j}^2 - O_{2,j}^2 - \text{DB}_j;$$
$$j = 1, 2, \dots, 12, \tag{5.10}$$

where
DB_j = diversion demand at Bowatenna in period j (10^6 m^3).
For Moragahakanda Reservoir:

$$S^2_{3,j+1} = S^2_{3,j} + I^2_{3,j} - E^2_{3,j} - R^2_{3,j} - O^2_{3,j} + R^2_{2,j} + O^2_{2,j};$$
$$j = 1, 2, \ldots, 12; \qquad (5.11)$$

$$R_{3,j} + O_{3,j} + \text{IE}_j \geq \text{DE}_j; \qquad j = 1, 2, \ldots, 12, \qquad (5.12)$$

where
IE_j = incremental inflow to Elahera diversion during period j (10^6 m^3), and
DE_j = diversion demand at Elahera in period j (10^6 m^3).
For the VRR subsystem:

$$Q_{v,j} + Q_{u,j} = Q^s_{p,j}; \qquad j = 1, 2, \ldots, 12, \qquad (5.13)$$

$Q_{v,j}$ = inflow entering VRR subsystem across interface point at Polgolla in period j (10^6 m^3).
For Victoria Reservoir:

$$S^3_{1,j+1} = S^3_{1,j} + I^3_{1,j} - E^3_{1,j} - R^3_{1,j} - O^3_{1,j} + Q_{v,j};$$
$$j = 1, 2, \ldots, 12. \qquad (5.14)$$

For Randenigala Reservoir:

$$S^3_{2,j+1} = S^3_{2,j} + I^3_{2,j} - E^3_{2,j} - R^3_{2,j} - O^3_{2,j} + R^3_{1,j} + O^3_{1,j};$$
$$j = 1, 2, \ldots, 12. \qquad (5.15)$$

For Rantembe Reservoir:

$$R^3_{3,j} = R^3_{2,j} + O^3_{2,j} + I^3_{3,j}; \qquad j = 1, 2, \ldots, 12, \qquad (5.16)$$

$$R^3_{3,j} + O^3_{3,j} + \text{IM}_j \geq \text{DM}_j; \qquad j = 1, 2, \ldots, 12, \qquad (5.17)$$

where
IM_j = incremental inflow to Minipe during period j (10^6 m^3), and
DM_j = diversion demand at Minipe in period j (10^6 m^3).
The following general equations apply to all three models:

$$S^N_{i,j+1} = S^N_{i,1}; \qquad j = 12; \qquad N = 1, 2, 3;$$
$$i \in I_N; \qquad I_1 = \{1, 3\}, \qquad (5.18)$$
$$I_2 = \{2, 3\}, \qquad I_3 = \{1, 2\};$$

$$O^N_{i,j} = R^N_{i,j} - \text{RMAX}^N_{i,j}; \qquad R^N_{i,j} \geq \text{RMAX}^N_{i,j};$$
$$N = 1, 2, 3; \qquad i = 1, 2, 3; \qquad j = 1, 2, \ldots, 12; \qquad (5.19)$$

$$R^N_{i,j} = \text{RMAX}^N_{i,j}; \qquad R^N_{i,j} \geq \text{RMAX}^N_{i,j};$$
$$N = 1, 2, 3; \qquad i = 1, 2, 3; \qquad j = 1, 2, \ldots, 12; \qquad (5.20)$$

and

$$O^N_{i,j} = 0; \qquad R^N_{i,j} \leq \text{RMAX}^N_{i,j}; \qquad N = 1, 2, 3; \qquad i = 1, 2, 3;$$
$$j = 1, 2, \ldots, 12;$$
$$S^N_{i,j+1} \leq \text{RMAX}^N_{i,j+1}; \qquad N = 1, 2, 3; \qquad i \in I_N;$$
$$j = 1, 2, \ldots, 12; \qquad (5.21)$$

$$\text{EL}^N_{i,j} = \text{SE}^N_i\left[\left(S^N_{i,j} + S^N_{i,j+1}\right)/2\right]; \qquad N = 1, 2, 3; \qquad i \in I_N;$$
$$j = 1, 2, \ldots, 12;$$
$$\text{UWL}^N_i; \qquad N = 1, 2, 3; \qquad i \notin I_N;$$
$$j = 1, 2, \ldots, 12; \qquad (5.22)$$

$$\text{DWL}^N_{i,j} = \max\left[\text{TWL}^N_i, \text{EL}^N_{i+1,j}\right]; N = 1, 2, 3; i = 1 \in I_N;$$
$$j = 1, 2, \ldots, 12;$$
$$\text{TWL}^N_i; \qquad N = 1, 2, 3; i = 1 \notin I_N;$$
$$j = 1, 2, \ldots, 12; \qquad (5.23)$$

where
UWL^N_i = upstream water level of (run-of-the-river) power plant i of the subsystem N (m),
$\text{DWL}^N_{i,j}$ = average downstream water level of power plant i in subsystem N during month j (m), and
TWL^N_i = normal tail water level of power plant i in subsystem N (m),
$\text{EL}^N_{i,j}$ = average water surface elevation of reservoir i in subsystem N during month j (m),
SE^N_i = represents the relationship between the water surfaces.
In the case of an objective function which minimizes the expected sum of squared deviations of water supply from demand, the demands are not considered as constraints. The results of the sequential optimization model obtained by using different diversion policies at Polgolla and Bowatenna are presented in Tables 5.12 and 5.13. These diversion policies consider different combinations of diversion capacity at Polgolla and the minimum release limits of Polgolla and Bowatenna. In this analysis, objective functions of maximization of expected energy generation and minimization of expected squared deviation of (supply–demand) were considered separately.

Results obtained by using the energy objective have outperformed those obtained by using the squared deviation objective.

5.2.8 Iterative upstream-moving decomposition: Mahaweli system

In this model, optimization of the system, which consists of three subsystems, is carried out using an iterative approach.

Table 5.12. *Results of the sequential optimization model (objective function: maximize energy generation)*

Alternative number	Polgolla minimum release (10^6 m^3)	Bowatenna minimum release (10^6 m^3)	Average annual energy generation (GWh)	Total annual firm energy (GWh)	Average annual water shortage at Minipe (10^6 m^3)	Average annual water shortage at Bowatenna (10^6 m^3)
Polgolla diversion policy: divert up to maximum of 75×10^6 m^3						
1	0	0	2853.6	625.0	97.8	0.5
2		10	2852.8	636.2	97.8	2.8
3		21	2847.1	650.8	97.8	26.0
4	11.2	0	2849.7	579.7	92.8	5.0
5		10	2849.8	594.0	92.8	12.3
6		21	2843.5	604.1	92.8	34.7
7	20	0	2859.1	557.8	84.7	16.8
8		10	2859.5	577.0	84.7	20.2
9		21	2854.4	593.6	84.7	47.5
10	30	0	2865.0	534.6	80.6	24.6
11		10	2864.8	551.9	80.6	34.7
12		20	2860.4	570.2	80.6	66.0
Polgolla diversion policy: divert up to maximum of 89×10^6 m^3						
13	0	0	2804.2	543.4	114.2	4.3
14		10	2804.9	551.4	114.2	10.3
15		21	2803.9	563.8	114.2	17.0
16	11.2	0	2814.9	528.2	102.9	14.8
17		10	2816.1	546.6	102.9	21.2
18		21	2817.6	565.6	102.9	30.1
19	20	0	2810.5	506.9	99.7	22.8
20		10	2812.4	522.0	99.7	32.0
21		21	2814.4	543.3	99.7	46.3
22	30	0	2815.2	516.5	89.3	36.3
23		10	2817.2	532.2	89.3	51.9
24		20	2813.5	551.8	89.3	67.7

As indicated in Figure 5.7, the iteration starts with the optimization of two downstream subsystems considering no inflows from Polgolla (i.e., considering only the incremental inflows into the downstream reservoirs) followed by two independent simulation runs. Historical monthly inflows and the operation policies derived in the optimization process are used in this simulation. Two time series of water shortages, one for each subsystem are thereby determined.

In the next step, the operation of the upstream subsystem consisting of Caledonia, Talawakelle, and Kotmale (CTK) Reservoirs is optimized. The shortages of two downstream subsystems determined in the previous step are considered as water demands for this system. Operation of the CTK subsystem is then simulated according to the formulated operation policies and historical inflows. This results in a time series of inflows at Polgolla Barrage. Diversions and downstream releases at Polgolla are determined according to several different diversion policies. The whole procedure is then repeated considering the new time series of diversions and spillages at Polgolla also as inflows to the two downstream subsystems. Iteration is continued until convergence to a constant system return is achieved. An average of four iterations is required to achieve convergence of these models.

Both objective functions of maximization of expected energy generation and minimization of expected squared deviation of (supply−demand) are considered separately.

The components of the mathematical formulation of the iterative optimization approach that are different from the sequential optimization approach (Section 5.2.7) are given below.

For the downstream subsystems, equations that correspond to Eq. (5.8) and Eq. (5.13) of the sequential optimization approach can be expressed as

$$Q_{u,j}(I) = D\left[Q_{p,j}^s(I-1)\right], \qquad I \geq 2; \quad j = 1,2,\ldots,12;$$
$$= 0 \qquad\qquad\qquad I = 1; \quad j = 1,2,\ldots,12; \tag{5.24}$$

Table 5.13. *Results of the sequential optimization model (objective function: minimize squared deviation of water supply from the demand)*

Alternative number	Polgolla minimum release (10^6 m^3)	Bowatenna minimum release (10^6 m^3)	Average annual energy generation (GWh)	Total annual firm energy (GWh)	Average annual water shortage at Minipe (10^6 m^3)	Average annual water shortage at Bowatenna (10^6 m^3)
Polgolla diversion policy: divert up to maximum of 75×10^6 m^3						
25	0	0	2778.7	633.8	122.8	4.3
26		10	2778.6	644.0	122.8	8.8
27		21	2783.1	660.6	122.8	29.3
28	11.2	0	2775.0	544.4	108.0	11.7
29		10	2776.0	554.3	108.0	19.2
30		21	2780.4	570.4	108.0	41.6
31	20	0	2783.3	558.9	103.3	19.3
32		10	2785.3	569.2	103.3	30.5
33		21	2790.9	584.0	103.3	57.8
34	30	0	2784.2	505.1	91.8	32.4
35		10	2786.7	521.0	91.8	47.0
36		20	2792.2	538.1	91.8	76.3
Polgolla diversion policy: divert up to maximum of 89×10^6 m^3						
37	0	0	2732.1	533.7	135.6	6.1
38		10	2733.2	544.4	135.6	10.3
39		21	2733.4	554.3	135.6	16.9
40	11.2	0	2737.7	523.9	128.6	12.0
41		10	2739.3	536.1	128.6	20.1
42		21	2740.3	546.6	128.6	30.6
43	20	0	2736.5	550.4	117.1	24.0
44		10	2738.5	561.3	117.1	34.3
45		21	2740.4	573.0	117.1	46.9
46	30	0	2735.1	538.6	106.6	35.6
47		10	2738.0	551.6	106.6	50.3
48		20	2740.7	563.7	106.6	69.0

$$Q_{u,j}(I) + Q_{v,j}(I) = Q_{p,j}^s(I-1), \qquad I \geq 2; \quad j = 1, 2, \dots, 12;$$
$$= 0, \qquad I = 1; \quad j = 1, 2, \dots, 12;$$
$$(5.25)$$

where

$Q_{p,j}^s(I-1)$ = inflow at Polgolla in period j obtained by simulating CTK subsystem according to its optimum operation policies during iteration $(I-1)$,

$Q_{u,j}(I)$ = inflow entering UBM subsystem across interface point at Polgolla in period j of iteration I (determined according to Polgolla diversion policy),

$D(.)$ = representation of diversion policy at Polgolla, and

$Q_{v,j}(I)$ = inflow entering VRR subsystem across interface point at Polgolla in period j of iteration I (determined according to Polgolla diversion policy).

For the upstream subsystem, the demand constraint that corresponds to Eq. (5.7) of the sequential optimization can be expressed as

$$Q_{p,j}(I) \geq SH_{u,j}^s(I) + SH_{v,j}^s(I) \quad \text{for all} \quad I, j = 1, 2, \dots, 12, \quad (5.26)$$

where $SH_{u,j}^s(I)$ and $SH_{v,j}^s(I)$ are the water shortages at period j obtained by simulating the operation of the UBM and VRR subsystems according to their optimal operation policies during the Ith iteration.

The same formulation except for the demand constraints is applicable when considering the objective function of minimization of the squared deviation of water supply from demand. The results of the iterative optimization obtained using several different diversion policies for the Polgolla diversion are presented in Table 5.14.

It is observed that the alternative solutions 53, 54, and 56 of the iterative optimization are not practically acceptable as they are associated with very high water shortages at Bowatenna. The results of a compromise programming analysis performed on the results of iterative and sequential approaches of Tables 5.12, 5.13, and 5.14 are presented in Table 5.15.

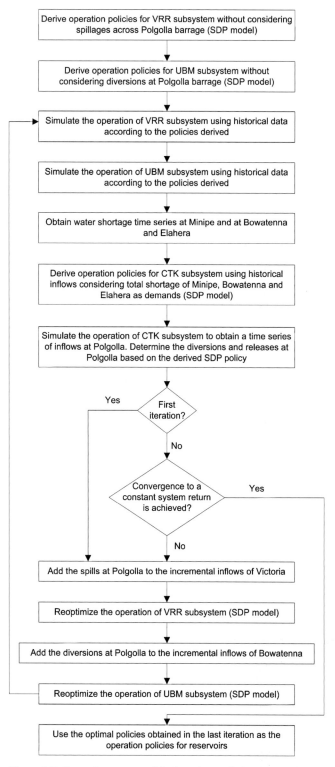

Figure 5.7 General structure of the iterative optimization model

policy which diverts water at Polgolla according to the average annual shortages of the downstream subsystems are found to be inferior to those corresponding to the diversion policy of Figure 5.14.

If the weight sets 4, 5, and 7 are excluded from consideration (since they do not properly represent the practical importance of the performance criteria of the Mahaweli system), alternatives 2 and 8 can be selected as the most satisfactory ones. Although alternative 8 slightly outperforms 2 in terms of average annual energy generation (an increase of 0.2%) and in terms of average annual water shortage at Minipe (a decrease of 13.4%), the high water shortage at Bowatenna (an increase of 620%) and the low firm energy generation (a decrease of 9.3%) make it inferior to alternative 2.

5.3 COMPOSITE RESERVOIR MODEL FORMULATION

Formulation of a hypothetical composite reservoir instead of the real multireservoir configuration is a convenient method to circumvent the "curse of dimensionality" of DP based operational optimization models. The composite representation of a serially linked two-reservoir system used by Kularathna (1992) is displayed in Figure 5.8. The fundamental idea behind the formulation of a hypothetical composite reservoir instead of the consideration of A and B reservoirs as individual units is to reduce the number of state variables and thereby reduce the computer memory requirement. Thus a larger part of the system can be handled in a single SDP model.

The composite reservoir concept can be presented by the following simplifications:

$$Q_c^j = Q_a^j + \beta \times Q_b^j; \qquad j = 1, 2, \ldots, N; \qquad (5.27)$$

$$S_c = S_a + S_b; \qquad j = 1, 2, \ldots, N, \qquad (5.28)$$

where
Q_c^j = inflow to composite reservoir in stage j (10^6 m^3),
Q_a^j, Q_b^j = inflows to reservoirs A and B in stage j (10^6 m^3),
S_c = active storage capacity of composite reservoirs (10^6 m^3),
S_a, S_b = active storage capacities of reservoirs A and B respectively (10^6 m^3),
N = number of stages, and
β = fraction of reservoir B's inflow assumed to be regulated by composite reservoir.

The justification of the above formulation is presented as follows. Since inflow to A is regulated by both reservoirs, the inflows to A are assumed to be completely passing through the

The results of Table 5.15 also indicate the suitability of an objective function which maximizes the expected annual energy generation to formulate operation policies for this particular system. The results obtained by using a diversion

Table 5.14. *Results of the iterative optimization model*

Alternative number	Polgolla minimum release (10^6 m^3)	Bowatenna minimum release (10^6 m^3)	Average annual energy generation (GWh)	Total annual firm energy (GWh)	Average annual water shortage at Minipe (10^6 m^3)	Average annual water shortage at Bowatenna (10^6 m^3)
Objective function: maximize energy generation						
Polgolla diversion according to Figure 5.14						
Maximum diversion $= 75 \times 10^6$ m^3/month						
Constraints: water demands/downstream shortages						
49	0.0	10.0	2858.7	589.9	103.0	2.5
50	11.2	10.0	2851.8	650.5	103.3	5.3
51	20.0	0.0	2844.3	618.2	98.3	8.0
52	20.0	10.0	2845.0	628.7	96.6	14.9
Polgolla diversion according to average annual shortages						
Maximum diversion $= 75 \times 10^6$ m^3/month						
Constraints: water demands/downstream shortages						
53	0.0	10.0	3003.6	806.2	51.5	175.2
54	11.2	10.0	2951.7	696.5	60.9	99.7
55	20.0	0.0	2924.4	601.9	56.5	45.9
56	20.0	10.0	2938.1	622.0	65.0	91.5
Objective function: minimize square deviation of (water supply – demand)						
Polgolla diversion according to Figure 5.14						
Maximum diversion $= 75 \times 10^6$ m^3/month						
57	0.0	10.0	2780.3	720.8	125.6	3.0
58	11.2	10.0	2778.2	569.6	118.8	11.3
59	20.0	0.0	2785.0	558.4	110.3	12.0
60	20.0	10.0	2784.5	579.4	110.3	20.4
Polgolla diversion according to average annual shortages						
Maximum diversion $= 75 \times 10^6$ m^3/month						
61	0.0	10.0	2796.4	725.4	109.6	33.5
62	11.2	10.0	2790.8	670.9	110.5	32.9
63	20.0	0.0	2784.9	522.9	107.4	19.7
64	20.0	10.0	2787.3	601.2	105.7	34.3

Table 5.15. *Results of the compromise programming analysis performed on the results of iterative and sequential optimization approaches*

Set	Weights				Alternatives ranked in positions 1, 2, and 3[a]
	Annual energy	Firm energy	Water shortage at Minipe	Water shortage at Bowatenna	
1	0.25	0.25	0.25	0.25	2, 3, 1
2	0.20	0.30	0.25	0.25	2, 3, 50
3	0.15	0.25	0.30	0.30	2, 52, 1
4	0.10	0.40	0.25	0.25	3, 50, 58
5	0.00	0.50	0.25	0.25	58, 3, 59
6	0.00	0.30	0.35	0.35	2, 52, 1
7	0.00	0.70	0.15	0.15	58, 54, 59
8	0.10	0.20	0.35	0.35	8, 7, 5

[a] Alternatives in this table are presented in Tables 5.12, 5.13 and 5.14

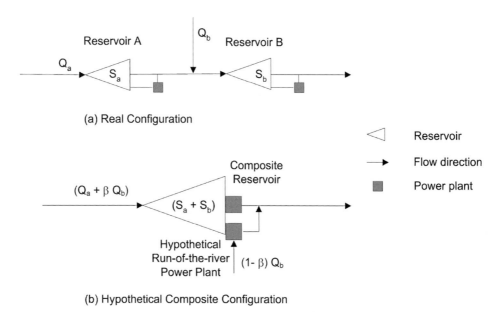

Figure 5.8 Composite representation of a serially linked two-reservoir system

composite reservoir as well. Incremental inflows to B are regulated only by the reservoir B. Therefore, in the composite formulation only a fraction of incremental inflows to B will be considered to pass through the composite reservoir. The fraction β is justified due to the reason that out of the total storage $(S_a + S_b)$ of the composite reservoir the inflow component of B is regulated only by a partial storage volume equal to the storage of reservoir B.

According to the above formulation, the total inflow to the composite reservoir will be $Q_a^j + \beta \times Q_b^j$. However, the real total inflow volume of A and B reservoirs is $Q_a^j + Q_b^j$. The leftover volume of water is therefore considered to be added to the downstream of the composite reservoir through a hypothetical hydropower plant, which has a generating head proportional to the head of the composite reservoir during the particular time period. This represents the fact that all the releases of reservoir B are passed through the power plant, subject to the limitations of the turbine capacities.

The A + B composite reservoir has to be formulated in such a way that it represents the performance of the real multireservoir subsystem fairly accurately. To achieve this similarity of the output, the performance of the assumed composite reservoir formulation is calibrated against that of the real configuration. The calibration is performed by formulating two optimization models. The first model considers the real multireservoir configuration of the reservoir system, while the second model uses the composite configuration. Optimal operation patterns obtained by the multireservoir formulation are compared with that obtained by the composite

reservoir optimization model formulation. A trial-and-error procedure is used to determine the parameters of the composite reservoir so as to obtain a performance similar to that of the multireservoir model. The model parameters include the inflow factor (β), the head factor (μ) (generating head of hypothetical power plant $= \mu \times$ head of composite reservoir) and the elevation–storage-area relationships of the composite reservoir. In the trial-and-error estimation of the parameters, the optimization of the composite reservoir operation is repeated by changing its parameters until the results of the composite optimization closely follow that of the multireservoir optimization. Comparison of the operation patterns is based on the monthly and annual energy/release plots of the two cases.

5.3.1 Analysis of the Mahaweli system based on three subsystems

The Mahaweli water resources system presented in Section 2.2.1 is used in this section. The operation policy analysis of the system initially requires an estimation of system water demands. These demands are to be estimated for an adequate time period upon which the operation policy analysis is to be performed. Historical monthly hydrological data are available for a period of 37 years (1949–85). Monthly time steps are considered throughout the analysis.

There are five major points in the system where water demands exist for the purpose of supplying the system irrigation areas. As shown in Figure 2.11, these diversion structures are located at Bowatenna, Elahera, Angamedilla, Minipe, and Kandakadu. Diversion water demands at these locations can be determined based on the irrigation water demands of the individual irrigation areas.

To mitigate computational burden, the Mahaweli system is simplified into three interconnected subsystems:

(a) Caledonia–Talawakelle–Kotmale (CTK) reservoir subsystem,

(b) Ukuwela–Bowatenna–Moragahakanda (UBM) subsystem,

(c) Victoria–Randenigala–Rantembe (VRR) subsystem.

The particular reason for identifying these three subsystems is that they reduce the number of interface points to a minimum of only one while forming computationally manageable subsystems. The Polgolla diversion structure acts as the interface point of these three subsystems. Hence the determination of the optimum diversion strategy for the Polgolla diversion structure is of utmost importance. It is also an important operational decision to be made in the actual system operation as well.

In this study, the operation of the upper Uma Oya Reservoir (Figure 2.11) was optimized independently of the other system components. The optimum release pattern obtained by simulating the operation of this reservoir was considered throughout the study as part of the incremental inflows to the Rantembe Reservoir.

5.3.2 Three-composite-reservoir IDP model

The effects of all three subsystems have to be considered jointly to determine the optimum diversion policy at Polgolla. Since consideration of the real multireservoir configuration is impractical due to the dimensionality of the problem, a composite representation of each subsystem has been used to circumvent the computational difficulties of the analysis. The approach converts the real multireservoir configuration into a three-reservoir system consisting of only three composite reservoirs interlinked at a common point. The common point in the three-composite-reservoir corresponds to the Polgolla Barrage.

Initially the three individual composite reservoirs have to be formulated and calibrated. Calibrations are based on the deterministic incremental dynamic programming (IDP) technique using 37 years of available monthly data. The parameters of each of the composite reservoirs are adjusted by a trial-and-error procedure until their optimal operation patterns yield similar results to those of the corresponding multireservoir optimizations. Two different objective functions: maximizing energy generation, and minimizing the squared deviation from the irrigation water demand, are used in separate calibration runs. The calibration results of the three composite reservoirs corresponding to squared deviation objective function are displayed in Figures 5.9–5.11.

Instead of optimizing the real multireservoir system, the resulting three-composite-reservoir configuration displayed in Figure 5.12 is considered for a deterministic analysis of the whole system. The aim of this analysis is to determine an optimal practically acceptable diversion policy for the Polgolla Barrage on distributing the inflow at Polgolla towards the two downstream subsystems. It is based on an IDP based optimization model for the three-composite-reservoir configuration having an objective function which minimizes the sum of the squared deviations of the water supplies from the diversion

(a)

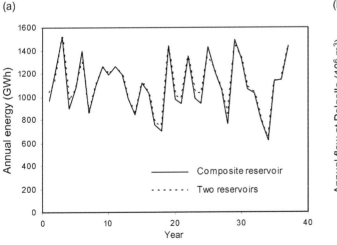

(b)

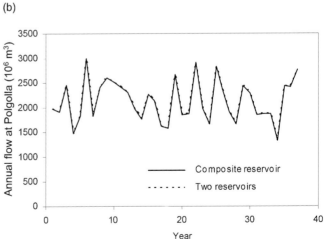

Figure 5.9 Calibration of Caledonia + Kotmale (C + K) composite reservoir (objective function: minimization of the squared deviation of water supply from the irrigation demand). (a) Comparison of annual energy. (b) Comparison of annual flows at Polgolla.

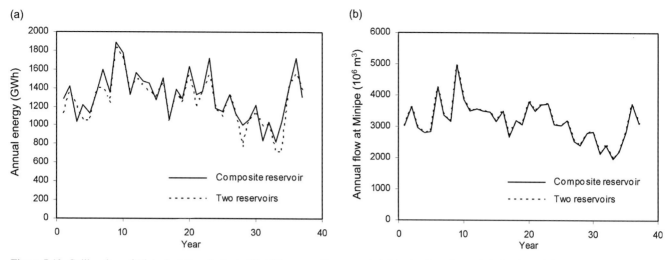

Figure 5.10 Calibration of Victoria + Randenigala (V + R) composite reservoir (objective function: minimization of the squared deviation of water supply from the irrigation demand). (a) Comparison of annual energy. (b) Comparison of annual flows at Minipe.

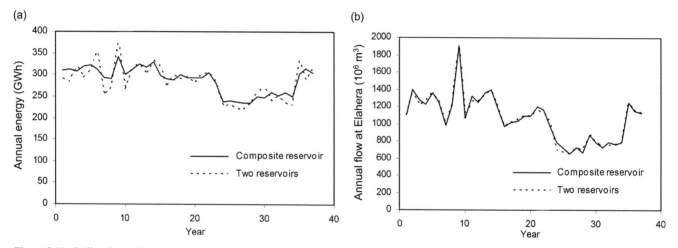

Figure 5.11 Calibration of Bowatenna + Moragahakanda (B + M) composite reservoir (objective function: minimization of the squared deviation of water supply from the irrigation demand). (a) Comparison of annual energy. (b) Comparison of annual flows at Elahera.

demands. The time span considered is the 37-year historical period from 1949 to 1985 on monthly time steps. The stages of the model are the time periods while the decisions comprise the monthly releases of the (composite) reservoirs, the diversion volume at the common interface which represents the Polgolla Barrage, and diversion volumes at Bowatenna, Elahera, and Minipe.

Three-composite-reservoir model formulation:
Objective function:

$$OF = \text{Minimize}\left\{\sum_{i=1}^{37}\sum_{j=1}^{12}TSD_{i,j}\right\}, \qquad (5.29)$$

where

$TSD_{i,j}$ = Sum of squared deviations of irrigation water supply from demand at Bowatenna, Elahera, and Minipe respectively in month j of year i.

$$TSD_{i,j} = (QB_{i,j} - DB_{i,j})^2 + (QE_{i,j} - DE_{i,j})^2 + (QM_{i,j} - DM_{i,j})^2;$$
$$i = 1, 2, \ldots, 37; \qquad j = 1, 2, \ldots, 12;$$

$QB_{i,j}$, $QE_{i,j}$, and $QM_{i,j}$ represent the volumes of water diverted at the Bowatenna Reservoir, Elahera diversion, and Minipe diversion respectively in month j of year i (10^6 m^3), and $DB_{i,j}$, $DE_{i,j}$, and $DM_{i,j}$ represent the diversion water demands at the

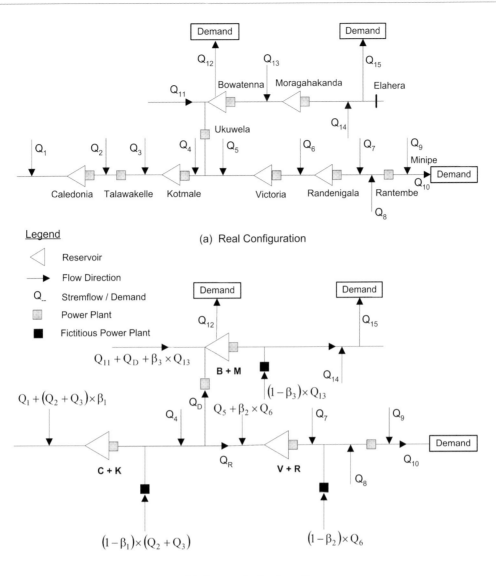

Figure 5.12 Real (a) and composite (b) configurations of the macrosystem

Bowatenna Reservoir, Elahera diversion, and Minipe diversion respectively in month j of year i ($10^6\,\text{m}^3$).

The continuity equations provide the basis for state transformation equations. Using the superscripts $N = 1$, 2, 3 respectively to represent the composite reservoirs of the three subsystems CTK, UBM, and VRR, the state transformation equations are presented in the following.

$$S_{i,j+1}^N = S_{i,j}^N + I_{i,j}^N - E_{i,j}^N - R_{i,j}^N - O_{i,j}^N;$$
$$N = 1, 3; \quad i = 1, 2, \ldots, 37; \quad j = 1, 2, \ldots, 12; \tag{5.30}$$

$$S_{i,j+1}^2 = S_{i,j}^2 + I_{i,j}^2 - E_{i,j}^2 - R_{i,j}^2 - O_{i,j}^2 - QB_{i,j};$$
$$N = 2; \quad i = 1, 2, \ldots, 37; \quad j = 1, 2, \ldots, 12; \tag{5.31}$$

$$S_{i,j+1}^N = S_{i+1,1}^N; \quad N = 1, 2, 3; \quad i = 1, 2, \ldots, 36; \quad j = 12; \tag{5.32}$$

$$O_{i,j}^N = R_{i,j}^N - RMAX_j^N; \quad R_{i,j}^N \geq RMAX_j^N;$$
$$N = 1, 2, 3; \quad i = 1, 2, \ldots, 37; \quad j = 1, 2, \ldots, 12; \tag{5.33}$$

$$R_{i,j}^N = RMAX_j^N; \quad R_{i,j}^N \geq RMAX_j^N; \quad N = 1, 2, 3;$$
$$i = 1, 2, \ldots, 37; \quad j = 1, 2, \ldots, 12; \tag{5.34}$$

and

$$O_{i,j}^N = 0.0; \quad R_{i,j}^N \leq RMAX_j^N; \quad S_{i,j+1}^N \leq SMAX_{j+1}^N;$$
$$N = 1, 2, 3; \quad i = 1, 2, \ldots, 37; \quad j = 1, 2, \ldots, 12; \tag{5.35}$$

$$QV_{i,j} = R_{i,j}^1 + O_{i,j}^1 + IH_{i,j}^1 + IP_{i,j} - QU_{i,j};$$
$$i = 1, 2, \ldots, 37; \qquad j = 1, 2, \ldots, 12; \tag{5.36}$$

$$I_{i,j}^1 = II_{i,j}^1; \qquad i = 1, 2, \ldots, 37; \qquad j = 1, 2, \ldots, 12; \tag{5.37}$$

$$I_{i,j}^2 = QU_{i,j} + II_{i,j}^2; \qquad i = 1, 2, \ldots, 37;$$
$$j = 1, 2, \ldots, 12; \tag{5.38}$$

$$I_{i,j}^3 = QV_{i,j} + II_{i,j}^3; \qquad i = 1, 2, \ldots, 37;$$
$$j = 1, 2, \ldots, 12; \tag{5.39}$$

$$FE_{i,j} = R_{i,j}^2 + O_{i,j}^2 + IH_{i,j}^2 + IE_{i,j}; \qquad i = 1, 2, \ldots, 37;$$
$$j = 1, 2, \ldots, 12; \tag{5.40}$$

$$FM_{i,j} = R_{i,j}^3 + O_{i,j}^3 + IH_{i,j}^3 + IM_{i,j}; \qquad i = 1, 2, \ldots, 37;$$
$$j = 1, 2, \ldots, 12; \tag{5.41}$$

$$QB_{i,j} \leq DB_{i,j}; \qquad i = 1, 2, \ldots, 37; \qquad j = 1, 2, \ldots, 12; \tag{5.42}$$

$$QE_{i,j} = \text{Min}\{FE_{i,j}, DE_{i,j}\}; \qquad i = 1, 2, \ldots, 37;$$
$$j = 1, 2, \ldots, 12; \tag{5.43}$$

$$QM_{i,j} = \text{Min}\{FM_{i,j}, DM_{i,j}\}; \qquad i = 1, 2, \ldots, 37;$$
$$j = 1, 2, \ldots, 12; \tag{5.44}$$

where

$S_{i,j}^N$ = storage of reservoir N at beginning of month j of year i $(10^6\,\text{m}^3)$,

$I_{i,j}^N$ = inflow to reservoir N during month j of year i $(10^6\,\text{m}^3)$,

$E_{i,j}^N$ = losses (mainly evaporation) from reservoir N during month j of year i $(10^6\,\text{m}^3)$,

$R_{i,j}^N$ = release from reservoir N during month j of year i $(10^6\,\text{m}^3)$,

$O_{i,j}^N$ = spill from reservoir N during month j of year i $(10^6\,\text{m}^3)$,

$RMAX_j^N$ = maximum release from reservoir N during month j $(10^6\,\text{m}^3)$,

$SMAX_j^N$ = maximum storage of reservoir N at beginning of month j $(10^6\,\text{m}^3)$,

$IP_{i,j}$ = incremental inflow to Polgolla Barrage during month j of year i $(10^6\,\text{m}^3)$,

$QU_{i,j}$ = volume of water diverted at Polgolla into UBM subsystem during month j of year i $(10^6\,\text{m}^3)$,

$QV_{i,j}$ = volume of water released at Polgolla into VRR subsystem during month j of year i $(10^6\,\text{m}^3)$,

$II_{i,j}^N$ = incremental inflow to composite reservoir N during month j of year i $(10^6\,\text{m}^3)$,

$IH_{i,j}^N$ = inflow to hypothetical power plant of reservoir N during month j of year i $(10^6\,\text{m}^3)$,

$IE_{i,j}$ = incremental inflows to Elahera diversion during month j of year i $(10^6\,\text{m}^3)$,

$FE_{i,j}$ = total inflow to Elahera diversion during month j of year i $(10^6\,\text{m}^3)$,

$IM_{i,j}$ = incremental inflows to Minipe diversion during month j of year i $(10^6\,\text{m}^3)$, and

$FM_{i,j}$ = total inflow to Minipe diversion during month j of year i $(10^6\,\text{m}^3)$.

Apart from the constraints of storage and release limits, the following constraints are also imposed:

$$QU_{i,j} \leq CAP, \tag{5.45}$$
$$TEP_{i,j} \geq FIRM, \tag{5.46}$$

where

CAP = maximum monthly diversion capacity of diversion tunnel at Polgolla $(10^6\,\text{m}^3)$,

FIRM = prespecified monthly firm energy value (GWh), and

$TEP_{i,j}$ = total energy production of system in month j of year i (GWh).

The above optimization model has a separable objective function consisting of 444 (12×37) components. Thus it can be solved using a DP formulation. The monthly time steps (j) are the stages, formulating a DP problem of 444 stages. The state variables are the storage volumes of the three composite reservoirs ($S_{i,j}^N; N = 1, 2, 3$). The decision variables of stage j are the volumes of water diverted at Polgolla (QU_j), Bowatenna (QB_j), Elahera (QE_j), and Minipe (QM_j) and the release decisions of each composite reservoir ($R_{i,j}^N; N = 1, 2, 3$). Due to the dimensionality of the problem, IDP is used to solve the model.

The diversion decision at Polgolla is incorporated into the model by treating it similarly to a state variable. This is in addition to the state of the system represented by the storage volumes of the three composite reservoirs. With each combination of storage states of the three reservoirs, three values for the diversion decision at Polgolla are considered. Thus, the imaginary corridor of this IDP model is formed by 81 (3^4) points that represent the states of the system to be accounted for at each stage.

The Bellman recursive equation for the IDP formulation of the model can be expressed as

$$F_{j+1}^*(S_{j+1}) = \text{Min}_{D_j}\Big\{TSD_j(S_j, S_{j+1}) + F_j^*(S_j)\Big\}, \tag{5.47}$$

$$S_j = \Big\{S_j^1, S_j^2, S_j^3\Big\},$$
$$D_j = \Big\{QU_j, QB_j, QE_j, QM_j, R_j^1, R_j^2, R_j^3\Big\}, \tag{5.48}$$
$$j = 1, 2, \ldots, 444,$$

Table 5.16. *Results of the three-composite-reservoir IDP model*

Diversion capacity at Polgolla ($10^6 \, \text{m}^3$/month)	Average annual energy generation (GWh)	Annual firm energy (GWh)	Average annual water shortage at Minipe ($10^6 \, \text{m}^3$)	Average annual water shortage at Bowatenna ($10^6 \, \text{m}^3$)	Average annual water shortage at Elahera ($10^6 \, \text{m}^3$)
60	2689.78	865.2	47.00	81.41	0.0
75	2663.84	926.4	45.22	66.75	0.0
89	2645.56	870.0	44.87	54.44	0.0
119	2627.05	876.0	44.61	51.11	0.0
149	2617.30	854.4	44.90	49.92	0.0

where

S_j = state of system at stage j,

$\text{TSD}_j(S_j, S_{j+1})$ = squared deviation of irrigation water supply from demands during stage j,

D_j = decisions associated with state transformation from S_j to S_{j+1}, and

$F^*_{j+1}(S_{j+1})$ = minimum total of objective function value from stage 1 to stage $j+1$, when state at stage $j+1$ is S_{j+1}.

Due to the large number of discrete state transformations that have to be considered at each stage, it is not possible to consider the entire 37-year period in a single optimization run. Instead, the 37-year period is divided into eleven 3-year periods and one 4-year period. It is assumed that the composite reservoirs are half-full at the beginning and at the end of each of these 12 periods. The model is run for five different values of maximum diversion capacities at Polgolla. Diversion capacities of 40%, 50%, 60%, 80%, and 100% of the capacity of the diversion tunnel are considered at this step. The aggregated results of the analysis are presented in Table 5.16.

An attempt is made to fit a regression formula between diversion and inflow at Polgolla based on diversions obtained from the model for several inflow series at Polgolla. However, fitting a regression formula to predict diversion at Polgolla using inflow to Polgolla is not possible based on the results obtained. Figure 5.13 presents monthly diversions at Polgolla obtained from the results of the three-composite-reservoir IDP model. These results correspond to the model run with a maximum diversion capacity of $75 \times 10^6 \, \text{m}^3$/month.

The wide variation of the diversion volume despite the large number of cases where the diversion volume reaches the upper limit refers to a poor correlation between the two variables: inflows and diversions at Polgolla. Thus, to determine the best diversion policy at the Polgolla Barrage, a diversion policy which has a close resemblance to the diversion pattern obtained by the three-composite-reservoir model is employed.

Figure 5.14 displays the diversion policy considered for further analysis.

Figure 5.14 indicates a minimum downstream release volume at Polgolla. The part of the available inflow that is in excess of this minimum release is to be diverted to the Amban Ganga basin. A maximum limit for this diversion is also specified. Any excess over the maximum possible diversion at Polgolla is to be spilled downstream into the VRR subsystem. In this diversion policy, the best values for the minimum release volume and the maximum limit on the diversion are determined by performing a sensitivity analysis using the three-composite-reservoir IDP model. In the sensitivity analysis, several different combinations of the two parameters are used to prespecify several independent diversion policies. With each of these diversion policies, the three-composite-reservoir IDP model is run using the available historical records. As the diversion volume is no longer a decision variable, the number of states to be considered in each stage of this model is 27 (3^3). Due to the reduced computational load, this model could be run by dividing the 37-year time series into three 9-year periods and one 10-year period. For each of these four periods, it is assumed that the composite reservoirs are half-full at the beginning and also at the end.

Table 5.17 presents results of the sensitivity analysis done with 15 different combinations of minimum release and diversion capacities at Polgolla. The best alternative combination is selected based on a multicriterion decision-making technique, compromise programming (CP). This CP analysis is done giving different sets of weight factors for the performance criteria, and maximum monthly diversion capacities of $75 \times 10^6 \, \text{m}^3$ and $89 \times 10^6 \, \text{m}^3$ at Polgolla resulted as the best. Therefore, in further analysis of the system using stochastic optimization models, maximum diversion capacities of only $75 \times 10^6 \, \text{m}^3$ and $89 \times 10^6 \, \text{m}^3$ are taken into account.

The results of the three-composite-reservoir IDP model have narrowed the range of operating options that can be used for diversion at Polgolla. With these rather narrow

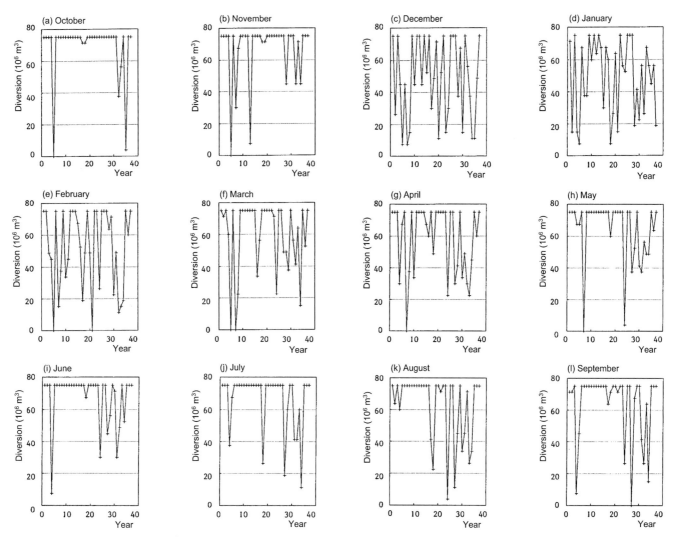

Figure 5.13 Monthly diversions at Polgolla based on the three-composite-reservoir IDP model

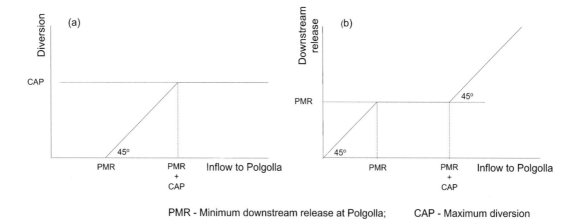

PMR - Minimum downstream release at Polgolla;   CAP - Maximum diversion

Figure 5.14 Polgolla diversion policy prespecified for the sensitivity analysis. (a) Diversion, (b) downstream release.

Table 5.17. *Sensitivity analysis results of the three-composite-reservoir IDP model*

Alternative number	Diversion capacity at Polgolla ($10^6\,m^3$/month)	Minimum downstream release at Polgolla ($10^6\,m^3$/month)	Average annual energy generated (GWh)	Annual firm energy (GWh)	Average annual water shortage at Minipe ($10^6\,m^3$)	Average annual water shortage at Bowatenna ($10^6\,m^3$)	Average annual water shortage at Elahera ($10^6\,m^3$)
1		0.0	2649.6	896.4	46.5	51.4	0.0
2	60.0	11.2[a]	2648.0	913.2	41.1	57.8	0.0
3		20.0	2651.2	901.2	38.6	61.4	0.0
4		0.0	2620.6	768.0	57.7	26.4	0.0
5	75.0	11.2	2622.3	849.6	47.2	38.2	0.0
6		20.0	2624.9	873.6	44.5	41.6	0.0
7		0.0	2595.8	703.2	67.2	9.2	0.0
8	89.0	11.2	2602.9	734.4	57.7	18.8	0.0
9		20.0	2610.4	783.6	50.6	26.8	0.0
10		0.0	2561.5	580.8	86.3	0.7	0.0
11	119.0	11.2	2576.3	579.6	77.6	3.0	0.0
12		20.0	2588.5	614.4	64.6	8.5	0.0
13		0.0	2520.6	516.0	111.6	0.1	0.0
14	149.0	11.2	2551.4	513.6	94.6	0.3	0.0
15		20.0	2558.6	558.0	80.2	1.1	0.0

[a] The minimum downstream release specified for the present operation of the Polgolla Barrage

operation patterns prespecified, the system is further analyzed in order to formulate operation policies of individual reservoirs. Two techniques, sequential optimization and iterative optimization, are presented in Sections 5.2.7 and 5.2.8, respectively. The principal idea of the sequential and iterative optimization approaches is to analyze the subsystems of the whole system separately, with the behavior at the interface point (Polgolla diversion) prespecified. For this purpose it is necessary to formulate SDP based optimization models for the individual multiunit subsystems. The assessment of the SDP model formulated for the VRR subsystem is presented in Section 3.2.

5.4 IMPLICIT STOCHASTIC DYNAMIC PROGRAMMING ANALYSIS

The dimensionality problems of using an explicit SDP approach can be avoided to some extent by implicitly incorporating the hydrological uncertainty. However, this may entail costs in terms of computational time as well as of the effectiveness of the resulting operation policies. This section derives an operation policy by the implicit stochastic approach and assesses its performance using the Victoria–Randenigala–Rantembe reservoir subsystem of the Mahaweli water resources system as the

case study area. A schematic diagram of the reservoir subsystem is displayed in Figure 5.15. In this analysis, the Rantembe Reservoir is treated as a run-of-the-river power plant and the operation policies are derived only for the Victoria and Randenigala Reservoirs. The analysis consists of generating several sets of streamflow data, followed by a deterministic optimization of the reservoir operation for each generated data set. The resulting optimum operation strategies are used in the derivation of operation rules using least-squares regression analysis.

5.4.1 Generation of synthetic streamflow data

Monthly streamflows to Victoria, Randenigala, Rantembe, and Minipe are generated using the "LAST" computer package developed by Lane and Frevert (1989) based on a 37-year-long historical streamflow data set. The upstream flows into the subsystem, namely, the flows across the Polgolla Barrage, are obtained by simulating the upstream reservoir subsystem according to its optimal operation policies. These flows are considered as deterministic flows in the implicit analysis. Being a biased hydrological data series estimated by a simulation model, the demand time series at Minipe is also considered as deterministic. A statistical analysis of the available historical data reveals that the incremental inflows at Victoria

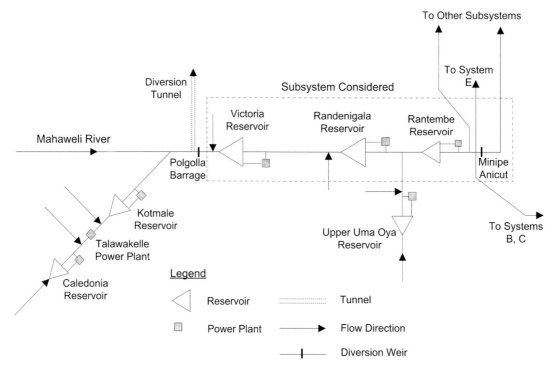

Figure 5.15 Schematic diagram of Victoria–Randenigala–Rantembe reservoir subsystem

and Randenigala are highly correlated. The correlation between flows at Rantembe and Minipe is also found to be high. In the data generation process, therefore, while considering Victoria and Rantembe as key stations, Randenigala and Minipe are analyzed as substations of Victoria and Rantembe respectively. For the annual to seasonal disaggregation, all four stations are considered simultaneously. The total length of generated data sequences is 74 years (two series each having a length of 37 years generated).

5.4.2 Optimization of system operation

Assuming each of the generated data sets and the original data set as deterministic streamflow sequences, the system operation is optimized in a deterministic environment using the incremental dynamic programming (IDP) technique. The state transformation equations that are the continuity equations of the reservoirs are the same as in Section 3.2. The constraints on minimum and maximum storage limits, minimum and maximum release volumes, and firm energy values as described in Section 3.2 also apply. The storage volumes of the two reservoirs represent the state of the system at each stage. Release volumes from the two reservoirs are the decisions that are to be made at each stage of the optimization process. The imaginary corridor that defines the limited state

space considered for this analysis consists of nine points (Section 2.2.1).

Two different objective functions, namely maximization of energy generation and minimization of squared deviation of the water supply from the irrigation demand are considered for the optimization. However, the downstream water demands are not considered as constraints because the solution becomes infeasible when this specific demand series is considered as constraints. Instead, feasible firm energy constraints, which are selected by trial-and-error, are imposed for both optimizations. These firm energy values are selected by gradually increasing the firm energy constraints of each reservoir until the solution becomes infeasible.

5.4.3 Regression analysis

Having formulated the deterministic optimum operation pattern for each streamflow sequence, a least-squares multiple regression analysis is performed (for the 12 months separately) to formulate an operation rule for the system. Thirty-five combinations of independent variables are considered in a preliminary regression analysis in order to determine the significant variables to formulate an operation policy. These independent variables include initial reservoir storages, inflows of reservoirs corresponding to the current

Table 5.18. *Combinations of independent variables selected for regression analysis of the implicit stochastic approach*

	Cross products/quadratic terms							Linear terms
	$S_{1,j}$	$S_{2,j}$	$Q_{1,j}$	$Q_{1,j-1}$	$Q_{2,j}$	$Q_{2,j-1}$	DM_j	
$S_{1,j}$	x	x	x	x	x	x	x	x
$S_{2,j}$		x	x	x	x	x	x	x
$Q_{1,j}$			x	x	x	x	x	x
$Q_{1,j-1}$				x	x	x	x	x
$Q_{2,j}$					x	x	x	x
$Q_{2,j-1}$						x	x	x
DM_j							x	x

and previous months, and irrigation water demand as linear terms. Their cross products and quadratic terms are also considered. Reservoir releases are considered as the dependent variables. Table 5.18 presents the independent variables for which the regression analysis is performed. From the results of the preliminary analysis, the combinations of independent variables that are found to be insignificant are removed and the analysis repeated with the remaining variables. The whole analysis is performed on the optimization results obtained by considering the two objective functions (maximum energy and minimum squared deviation) separately. This results in the operation rules expressed by the regression equations.

For Table 5.18:

$S_{i,j}$ = storage of reservoir i at beginning of month j (10^6 m^3),

$Q_{i,j}$ = inflow to reservoir i during month j (10^6 m^3), and

DM_j = difference between irrigation demand at Minipe and unregulated incremental inflows to Minipe during month j (10^6 m^3).

The operation rules derived by using an objective function of minimization of the squared deviation of the irrigation water supply from the demand can be expressed as

$$R_{1,j} = A_{1,j} \times S_{1,j} + A_{2,j} \times S_{2,j} + A_{3,j} \times DM_j + A_{4,j} \times (S_{1,j})^2$$
$$+ A_{5,j} \times (S_{2,j})^2 + A_{6,j} \times S_{2,j} \times Q_{1,j} + A_{7,j} \times S_{2,j} \quad (5.49)$$
$$\times DM_j + A_{8,j}; \qquad j = 1, 2, \ldots, 12;$$

$$R_{2,j} = B_{1,j} \times S_{2,j} + B_{2,j} \times (S_{2,j})^2 + B_{3,j} \times DM_j + B_{4,j} \times Q_{1,j}$$
$$+ B_{5,j} \times Q_{1,j-1} + B_{6,j} \times Q_{2,j} + B_{7,j} \times (Q_{2,j})^2 + B_{8,j} \quad (5.50)$$
$$\times (DM_j)^2 + B_{9,j}; \qquad j = 1, 2, \ldots, 12;$$

where

$R_{i,j}$ = release from reservoir i during month j (10^6 m^3) ($i = 1$ and 2 indicate Victoria and Randenigala Reservoirs respectively),

$A_{i,j}$; $i = 1, 2, \ldots, 8$; $j = 1, 2, \ldots, 12$ and $B_{i,j}$; $i = 1, 2, \ldots, 9$; $j = 1, 2, \ldots, 12$ are regression coefficients.

With the use of an objective function of maximization of energy generation, the following regression equations have been obtained:

$$R_{1,j} = C_{1,j} \times S_{1,j} + C_{2,j} \times Q_{1,j} + C_{3,j} \times Q_{1,j-1} + C_{4,j} \times DM_j$$
$$+ C_{5,j} \times (S_{1,j})^2 + C_{6,j} \times S_{1,j} \times Q_{1,j} + C_{7,j} \times S_{1,j}$$
$$\times Q_{1,j-1} + C_{8,j} \times Q_{1,j} \times DM_j + C_{9,j}; \qquad j = 1, 2, \ldots, 12;$$
$$(5.51)$$

$$R_{2,j} = D_{1,j} \times S_{1,j} + D_{2,j} \times Q_{1,j} + D_{3,j} \times Q_{2,j} + D_{4,j} \times DM_j$$
$$+ D_{5,j} \times (S_{1,j})^2 + D_{6,j} \times (Q_{1,j})^2 + D_{7,j} \times S_{1,j} \times Q_{1,j}$$
$$+ D_{8,j} \times S_{1,j} \times Q_{2,j} + D_{9,j} \times Q_{1,j} \times DM_j \quad (5.52)$$
$$+ D_{10,j}; \qquad j = 1, 2, \ldots, 12;$$

$C_{i,j}$; $i = 1, 2, \ldots, 9$; $j = 1, 2, \ldots, 12$ and $D_{i,j}$; $i = 1, 2, \ldots, 10$; $j = 1, 2, \ldots, 12$ are regression coefficients.

The regression coefficients of Eq. (5.49) to Eq. (5.52) are determined with the corresponding "coefficients of determination." According to the operation rules of Eq. (5.49) to Eq. (5.52), the system operation was simulated. The simulated system performance obtained by these implicit SDP based operation rules is compared with the simulation performed according to the explicit SDP based operation policies and also with the deterministic optimum operation and the historical operation in Table 5.19.

As the table shows, the IDP models indicate the upper bounds on the objective achievements for a particular historical data set. In the case of the squared deviation objective, objective achievement is indirectly indicated by firm energy generation. An explicit indication in terms of annual energy generation is made in the case of the energy objective. Table 5.19 shows that the explicit SDP based operation policy formulated by using the energy objective (alternative (5)) outperforms the implicit SDP based operations (1) and (4). Although (1) is preferable in terms of the probability of failure, (5) outranks (1) when considering the other three performance criteria, specially firm energy generation. It can be seen that the historical operation obtained by simulating the system using the present rule curves indicates lower annual energy and firm energy values, although the historical operation is slightly better in terms of the water shortage and the probability of failure months.

It can be noted that the model inaccuracies induced by the implicit stochastic approach are quite significant. These inaccuracies accrue in the first instance during the data generation process. Subsequent regression analysis increases the level of inaccuracy. These inaccuracies could be further enhanced in the case of a more complex reservoir system.

Table 5.19. *Summary comparison of performance of implicit SDP based operation with that of explicit SDP based operation, deterministic optimum, and historical operation*

Alternative		Average annual energy (GWh)	Annual firm energy (GWh)	Average annual shortage at Minipe (10^6 m^3)	Probability of failure months (%)
	Objective function: minimize squared deviation				
1	Implicit SDP	1232.9	56.4	94.2	5.18
2	Explicit SDP	1203.0	144.7	139.3	9.01
3	IDP	1257.3	248.4	102.2	16.20
	Objective function: maximize energy generation				
4	Implicit SDP	1088.6	52.7	198.9	12.60
5	Explicit SDP	1283.8	158.4	87.2	5.40
6	IDP	1427.5	164.4	548.0	40.30
7	Historical operation	1258.0	102.8	82.6	5.18

5.5 DISAGGREGATION/AGGREGATION TECHNIQUES BASED ON DYNAMIC PROGRAMMING

5.5.1 Disaggregation of composite operation policies

Section 5.3 demonstrates the usefulness of a hypothetical composite reservoir formulation to mitigate dimensionality problems in analyzing multireservoir systems. Although the composite reservoir approach makes the analysis computationally manageable, it implies an operation policy for the hypothetical composite reservoir(s) only. Such an operation policy would be of very little use unless it is disaggregated to operation policies of the real individual reservoirs.

Three different approaches for the disaggregation of composite operation policies are proposed. Their applicability is tested by applying the techniques for the case of Victoria–Randenigala–Rantembe reservoir subsystem of the Mahaweli water resources system. In order to incorporate the stochastic nature of the inflows explicitly, the composite reservoir formulated in place of the Victoria and Randenigala Reservoirs is optimized using an explicit SDP model. The small downstream reservoir at Rantembe is considered as a run-of-the-river power plant. Two different objective functions are considered. Maximization of expected energy generation subject to the downstream water demand constraints is one of them. The second formulation is to minimize the expected sum of squared deviations of water supply from the irrigation demand at Minipe. The V + R composite reservoir, which has been calibrated in Section 5.3, is used here in a stochastic context. Simulated performances of the V + R composite reservoir are compared with the performances of the realistic V&R two-reservoir system in Table 5.20.

The three disaggregation approaches proposed for disaggregating composite policies are:

(a) statistical disaggregation of composite policies,
(b) disaggregation by an optimization/simulation based approach,
(c) use of a deterministic optimization model in each time interval of the operation.

STATISTICAL DISAGGREGATION OF COMPOSITE POLICIES

The statistical disaggregation model of Lane and Frevert (1989) generates seasonal flows by disaggregating annual flows to seasonal values. In this approach, key stations (stations of major importance) are used to generate key and substation flows, preserving the intercorrelations between key and substations. This model preserves serial- and cross-correlations between variables on an annual as well as on a seasonal basis.

The LAST statistical disaggregation package originally formulated for hydrological data generation is applied for reservoir operation in this work. It is used to determine the optimum operation of individual reservoirs based on the operation of a hypothetical composite reservoir. The composite reservoir can be considered as a key station, while the individual reservoirs correspond to the substations.

Simulating the operation of the composite reservoir using historical streamflow data according to the SDP based policy, the optimum operation pattern for the composite reservoir is obtained (a sequence of monthly storage volumes during the period for which the historical data are available). This optimum operation pattern, together with the composite inflows, is treated as the key station data in the disaggregation approach.

Table 5.20. *Comparison of the simulated performance of Victoria + Randenigala (V + R) composite reservoir with that of the real Victoria and Randenigala (V&R) two-reservoir system*

Objective function/constraints	Configuration	Average annual energy (GWh)	Annual firm energy (GWh)	Average annual shortage at Minipe (10^6 m^3)	Probability of failure[a] months (%)
Maximize energy; demands at Minipe; firm energy constraints	V + R composite	1280.9	75.0	106.3	5.6
	V & R two-reservoir	1263.6	161.9	104.1	5.4
Minimize squared deviation of supply − demand at Minipe; firm energy constraints	V + R composite	1271.9	62.5	107.3	5.6
	V & R two-reservoir	1203.0	144.7	139.3	9.0

[a] Failure to satisfy the irrigation water demands

The actual operation patterns of the individual reservoirs that are considered as substations are required to implement the disaggregation approach. As an initial estimate, the historical operation patterns of the two reservoirs are used for this purpose. The pattern is obtained by simulating the reservoir subsystem using historical data according to the present rule curves. Monthly operational data (storage and inflow volumes) of these key stations and substations are then generated using the LAST disaggregation model. The statistics of the original and generated data for the two objective functions considered in the composite formulation are estimated. These generated data are used in a multiple linear regression analysis in order to formulate operation policies for the two individual reservoirs. It is found that the use of only linear terms as independent variables results in satisfactory values for R^2 (coefficient of determination).

The operation rules derived by the regression analysis can be presented as follows:

$$S_{1,j+1} = A_{1,j} \times S_{c,j+1} + A_{2,j} \times Q_{1,j} + A_{3,j} \times S_{1,j} + A_{4,j} \times Q_{2,j}$$
$$+ A_{5,j} \times S_{2,j} + A_{6,j}; \qquad j = 1, 2, \ldots, 12; \qquad (5.53)$$

$$S_{2,j+1} = B_{1,j} \times S_{c,j+1} + B_{2,j} \times Q_{1,j} + B_{3,j} \times S_{1,j} + B_{4,j} \times Q_{2,j}$$
$$+ B_{5,j} \times S_{2,j} + B_{6,j}; \qquad j = 1, 2, \ldots, 12; \qquad (5.54)$$

where

$S_{i,j}$ = storage volume of reservoir i at beginning of month j (10^6 m^3) (c refers to composite reservoir)

$Q_{i,j}$ = inflow to reservoir i during month j (10^6 m^3), and $A_{i,j}$ and $B_{i,j}$; $i = 1, 2, \ldots, 6$; $j = 1, 2, \ldots, 12$, are regression coefficients.

Regression coefficients of the above equations corresponding to the two objective functions can be obtained.

The performance of this methodology has been tested by simulating the performance of the V&R system according to the operation rules indicated by the above equations. The results of the analysis are presented in Table 5.21.

DISAGGREGATION OF COMPOSITE POLICIES BY OPTIMIZATION/SIMULATION BASED APPROACH

The aim of this approach is to determine the operation policies of individual reservoirs in such a way that they will reproduce, upon simulation, the simulated optimal operation pattern of a hypothetical composite reservoir.

The model formulation and the selection of simulated composite outputs to be reproduced in the real operation depend on the particular system configuration. In this study, however, the basis of the disaggregation approach is the reproduction of monthly flows at Minipe that were obtained using the composite formulation.

However, a two-reservoir system can be easily analyzed using the two-reservoir SDP models developed for this study (Section 3.2). In addition, a two-reservoir system with downstream demands can be analyzed without difficulty using even an iterative model formulation. In such a situation, determination of composite-reservoir-based flows at Minipe would not be necessary as the actual downstream demands at Minipe can be used directly. Nevertheless, the two-reservoir V&R system is selected in this analysis for the purpose of demonstrating the applicability of the proposed approach.

The reproduction of V + R composite flows in the realistic V&R case is attempted by analyzing the V&R two-reservoir configuration by an iterative optimization model. Minimization of the expected sum of the squared deviation of water flow at Minipe from that obtained by the composite formulation is the objective function. The iterative optimization is initiated with the optimization of the downstream Randenigala Reservoir followed by a simulation run of the same reservoir according to the optimum policy formulated in the optimization process. Simulated monthly water shortages at Minipe are then considered as water demands from the upstream reservoir Victoria, which is optimized next using a squared deviation objective function. Simulation of the upstream reservoir according to the optimum policy

Table 5.21. *Comparison of the results of the composite-policy-disaggregation approach*

Disaggregation approach	Objective function	Average annual energy (GWh)	Annual firm energy (GWh)	Average annual shortage at Minipe (10^6 m^3)	Probability of failure[a] months (%)
Statistical	Max. energy	1248.0	167.6	94.1	5.4
	Min. sq. deviation	1252.3	157.9	91.5	5.2
Optimization and simulation	Max. energy	1217.8	165.0	125.9	7.7
	Min. sq. deviation	1217.7	171.7	124.5	7.7
Use of a single-time-step optimization	Max. energy	Deterministic objective function (1)			
		1185.8	125.8	205.5	13.9
		Deterministic objective function (2)			
		1201.1	123.8	199.6	13.3
	Min. sq. deviation	Deterministic objective function (1)			
		1179.3	129.6	216.1	14.9
		Deterministic objective function (2)			
		1195.3	135.8	198.5	13.3

[a] Failure to satisfy the irrigation demands

formulated in the optimization results in a flow series into the downstream Randenigala Reservoir. With this new inflow series, the process is repeated until convergence to a constant system return is obtained. Corresponding operation policies of the individual reservoirs with which the convergence is obtained are considered as disaggregated operation policies for individual reservoirs.

The system operation according to the operation policies derived using this approach is simulated. A summary of the simulation results is presented in Table 5.21.

USE OF SINGLE-TIME-STEP OPTIMIZATION MODEL TO DISAGGREGATE COMPOSITE POLICY

This approach optimizes the operation strategy of the system for each time step, subject to the broad operation policy constraint set by the composite configuration. As in the previous stochastic models, it is assumed that a perfect streamflow forecast is available. The applicability of this approach is tested by coupling an elementary single-time-step optimization model to a simulation model that uses the composite policy as the basis for simulation.

The deterministic optimization model formulation is presented in the following. Optimization for each of the fixed monthly time intervals (j) is attempted by using two different objective functions. The elementary optimization procedure used here is to evaluate all feasible discrete combinations of decisions for each month separately. For each month of the total simulation period of 37 years:

$$(1) \quad \text{Max } Z_j = \sum_{i=1}^{2} (\text{TEP}_{i,j}); \qquad j = 1, 2, \ldots, 12; \qquad (5.55)$$

$$(2) \quad \text{Min } Z_j = \left[\left(\frac{S_{1,j+1}^{\max} - S_{1,j+1}}{S_{1,j+1}^{\max} - S_{1,j+1}^{\min}} \right)^2 + \left(\frac{S_{2,j+1}^{\max} - S_{2,j+1}}{S_{2,j+1}^{\max} - S_{2,j+1}^{\min}} \right)^2 \right]^{1/2}; \quad j = 1, 2, \ldots, 12; \qquad (5.56)$$

where

$\text{TEP}_{i,j}$ = energy generation of reservoir i during month j (GWh),

$S_{i,j+1}$ = storage volume of reservoir i at beginning of month $j+1$ (10^6 m^3),

$S_{i,j+1}^{\max}$ = maximum storage capacity of reservoir i at beginning of month $j+1$ (10^6 m^3), and

$S_{i,j+1}^{\min}$ = minimum storage capacity of reservoir i at beginning of month $j+1$ (10^6 m^3);

subject to:

$$\left| S_{c,j+1} - (S_{1,j+1} + S_{2,j+1}) \right| \leq \delta; \qquad j = 1, 2, \ldots, 12; \quad (5.57)$$

$$\text{TEP}_{i,j} \geq \text{FIRM}_{i,j}; \qquad j = 1, 2, \ldots, 12; \quad i = 1, 2; \quad (5.58)$$

where

δ = allowable deviation from prespecified composite policy (10^6 m^3),

$S_{c,j+1}$ = storage volume of composite reservoir at beginning of month $j+1$ (10^6 m^3) (specified by the composite policy),

$\text{TEP}_{i,j}$ = energy generation at reservoir i during month j (GWh), and

$\text{FIRM}_{i,j}$ = firm energy generation of reservoir i during month j (GWh).

In addition, the constraints on reservoir storage and release volumes as well as the continuity equations presented in Eq. (3.10) to Eq. (3.17) also apply. Composite policies derived by the two different model formulations described in Section 5.5.1 are used separately in the analysis. Results of the simulations performed using this technique are also presented in Table 5.21.

Table 5.21 shows that the statistical disaggregation approach is preferable over the other two. But the statistical disaggregation approach of this study is based on the historical operation pattern obtained by simulating the reservoir system according to its present operation rule curves. These rule curves are established after detailed simulations of the system. They may also be indicating a "near optimal" operation pattern. This precludes the possibility of selecting the statistical disaggregation approach as the most suitable one, since the superior results of it may be due to the biased "historical operation." The results of the single-time-step optimization can be described as unacceptable. However, this approach may be improved if the approximate operational behavior of the individual reservoirs is considered as guidance in the disaggregation approach. The possibility of performing deterministic optimization over a longer time span may also improve the resulting performance. By comparing the results obtained by considering the real two-reservoir configuration (Table 5.20) with those in Table 5.21, it can be concluded that the approach based on optimization and simulation is practically acceptable, as it yields fairly good results based on an uncomplicated analysis.

The iterative solutions and system decomposition/iteration are more viable than the composite reservoir, which needs a lot of calibration, and then disaggregation of the policy. Judgement and comparison of these two main approaches to overcome the dimensionality problem are highly desirable for practical purposes.

6 Optimal reservoir operation for flood control

Reservoir systems are operated in somewhat uncertain environments. The uncertainty is mainly due to forecasting of expected rainfall caused by events such as typhoons and resulting river inflows. The problem of optimizing real-time short-time on-line operation of complex reservoir systems under uncertainty is very difficult particularly during extreme events like floods or droughts. Practically no general solution is available for this type of problem. Thus, it is important to fully appreciate the problem of real-time reservoir operation under uncertainty before expedient methods for its solution can be developed. This section describes how far DP based operation can be used in the phase of short-term "emergency" operation or how far this type of short-term operation can be embedded in DP or SDP based rules.

Several operational modes can be employed for a water resources system. For example, in real-time on-line operation of a multipurpose reservoir, the importance of a purpose or particular demand may vary either periodically with the annual cycle or randomly due to the occurrence of floods. Real-time on-line operation may be defined as an interaction between the operator and the system while the operation is being executed, and the response time is critical within a definite time step. Consequently, real-time on-line operation may imply operational mode changes to respond to this critical situation. One of the most challenging decisions inherent in the operation of a reservoir is to decide when to change release policy and allocate storage space, for example for the purpose of flood control instead of for conservation storage, and vice versa. This reservoir management problem is frequently encountered in climatic zones dominated by a marked monsoon or typhoon-borne rainy season and a subsequent long dry season.

Reservoir operations are normally based upon fixed long-term operation rules. A sophisticated operation like a SDP based reservoir operation policy with its inherent slow response characteristics can hardly account for the swift variability of hydrological conditions like sudden onset floods. A long-term policy is rigid concerning short-term operation.

The decision whether or not and when to shift the operational mode from long-term operation to short-term operation and vice versa is quite crucial. A prolonged mismatch between the required and the actual reservoir storage level and release implies an opportunity loss of possibly considerable proportion.

It is very important to have a premeditated rule about when to switch the operational mode. The practice based on simulation-derived operational rule curves contains the provision to lower the reservoir water level during the typhoon/high flow season. This may result in reduced storage volumes if no major inflow events occur. To avoid this type of loss, as well as to minimize flood damage, the short-term emergency mode operation should be embedded into an optimization based long-term policy.

6.1 FEITSUI RESERVOIR PROJECT IN TAIWAN

The Feitsui Reservoir Project in Taiwan shown in Figure 6.1 has a catchment area of $303\,km^2$. The Feitsui Reservoir of capacity $406 \times 10^6\,m^3$ is located on Peishih Creek, a tributary of the Hsintien River, 30 km southeast of Taipei. The primary purpose of the Feitsui Reservoir is water supply for 4 million people living in the Taipei area, while hydro-energy generation by a power plant of 70 MW installed generating capacity is its secondary purpose. It is expected that with the help of this reservoir the estimated water demand of the Taipei metropolitan area can adequately be met till 2030.

The current long-term operating strategies of the Feitsui Reservoir (provided by Sinotech Engineering Consultants Inc., 1985) are based on rule curves giving the state of the reservoir as a function of time for an annual cycle. These rule curves are defined based on 10-day-long time steps. They have been developed by trial and error methods and tested through simulations based on historical and synthetic streamflow series.

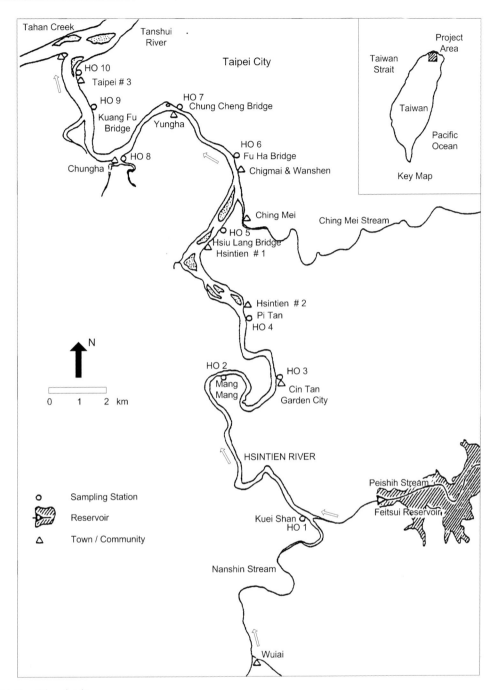

Figure 6.1 The Hsintien River basin

Yet these rule curves are only guides, and will not necessarily ensure the most rational utilization of water resources. Thus, the problem of how to utilize the water resources more efficiently can still be regarded as an unsolved task.

Though flood control is not an explicit purpose of the Feitsui Reservoir, it is taken into account in the reservoir operation during the typhoon season. Operating the system safely in the typhoon season is a challenging task. A procedure for switching the operational mode between short-term and long-term operations is required during a typhoon. The current short-term operation policy of the Feitsui Reservoir is still based on the upper limit of the above-mentioned rule curves. Certainly, it cannot guarantee optimal operation in the case of actual typhoon events.

6.2 OPERATIONAL MODE SWITCH SYSTEM BETWEEN LONG-TERM AND SHORT-TERM OPERATION

An operational mode switch (OMS) system developed by Huang (1989) for the Feitsui Reservoir determines whether or not and when to shift the operation back and forth between long-term "normal mode" and short-term "emergency mode" during a typhoon attack. Further, it considers stochasticity of inflows and streamflow forecasts in the development of long-term/short-term operation policies. The schematic representation of the method is shown in Figure 6.2 in which the OMS acts as a bridge to interlink the long- and short-term operations.

The flow chart of the OMS model for on-line reservoir operation is given in Figure 6.3. As shown in the figure, the long-term reservoir operation is based on an operation policy which is derived by SDP.

During the flood season in Taiwan, the streamflow (base flow) approaches stationarity because the inherent disturbance in the environment is not significant. Hence, the hydrological models such as Box–Jenkins models are applicable for long-term forecasts. However, due to the steep topographical slope and small catchment areas in Taiwan with sudden large inputs of rainfall from extreme events, Box–Jenkins models are not appropriate for short-term streamflow forecasts. Instead, other hydrological models such as rainfall–runoff models are preferred. The OMS model presented is associated with typhoon events where the base flow is not a significant proportion of the total flow in the river during the periods of rainfall. The streamflow series are extraordinarily influenced by noise. Hence, the attention of the OMS model is directed principally at forecasting river discharges with a rainfall–runoff model rather than with Box–Jenkins models.

6.3 DEVELOPMENT OF SDP MODEL FOR LONG-TERM OPERATION

The explicit stochastic dynamic programming (SDP) model is used for the derivation of long-term operation policies for the Feitsui Reservoir system.

6.3.1 Objective function for long-term operation

The primary purpose of the Feitsui Reservoir is water supply while hydro-energy generation is a secondary purpose. To utilize the available water resources efficiently, the water demand is set as a constraint while the maximization of the expected annual hydropower generation is regarded as the objective function. Hence, the objective function of the system is

$$OF = \text{Maximize } \xi \left[\sum_{j=1}^{T} EP_j \right], \qquad (6.1)$$

where

ξ = expectation operator,

EP_j = hydro-energy generation in period j, and

T = number of periods ($T = 36$ for a 10-day basis within one year).

Reservoir storage and inflow are state variables.

Based on different state variables and inflow transition probability in SDP, i.e., either current inflow or previous inflow as state variable to decide the operation policy, and either conditional or unconditional probabilities of inflow to obtain the steady-state policy, four different SDP models can be formulated. Let j refer to the within-year period and n to the total number of periods considered up to the actual stage. The indices S_j, Q_j, R_j, and S_{j+1} denote the states of initial storage, inflow, and release during the time step, and final storage (initial storage for the next time step), respectively. $P_{Q(j+1)|Q(j)}$ is the probability of inflow Q_{j+1} in period $j+1$, when the inflow in period j equals Q_j. The maximum value of energy production up to period n associated with the state variable values S_j and Q_j is f_j^n. Figure 6.4 indicates the relationship between these indices.

Accordingly, the recursive backward-moving dynamic relationships corresponding to different types of SDP can be expressed as follows:

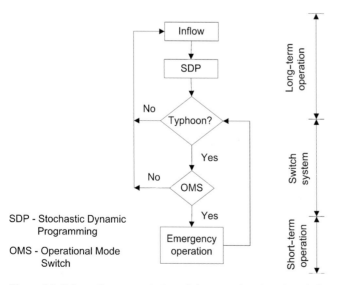

SDP - Stochastic Dynamic Programming

OMS - Operational Mode Switch

Figure 6.2 Schematic representation of the operational mode switch system

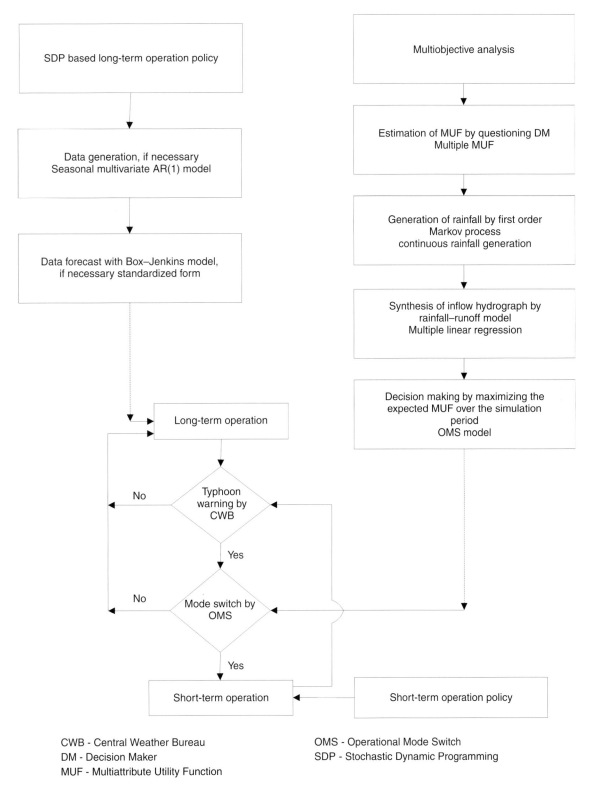

Figure 6.3 Flow chart of the OMS model for on-line reservoir operation

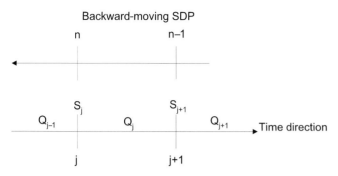

Figure 6.4 Relationship between variables of SDP

Type 1

$$f_j^n(S_j, Q_j) = \underset{S_{j+1}}{\text{Maximize}} \left[\text{EP}_j + \sum_{Q_{j+1}} P_{Q(j+1)|Q(j)} \times f_{j+1}^{n-1}(S_{j+1}, Q_{j+1}) \right].$$

(6.2)

Type 1 assumes that the probability of inflow is conditional, i.e., correlation between two consecutive inflows exists, and the current inflows are known perfectly already while deriving the steady-state operaton policy based on initial reservoir storage, current inflow, and optimal final reservoir storage target.

Type 2

$$f_j^n(S_j, Q_j) = \underset{S_{j+1}}{\text{Maximize}} \left[\text{EP}_j + \sum_{Q_{j+1}} P_{Q(j+1)} \times f_{j+1}^{n-1}(S_{j+1}, Q_{j+1}) \right].$$

(6.3)

Type 2 assumes unconditional probability of inflows. That is, there is no correlation between two consecutive inflows. In the meantime, perfect knowledge of the current inflows is available while deriving the steady-state operation policy in terms of initial reservoir storage, current inflow, and optimal final storage volume target.

Type 3

$$f_j^n(S_j, Q_{j-1}) = \underset{R_j}{\text{Maximize}} \sum_{Q_j} P_{Q(j)|Q(j-1)} \left[\text{EP}_j + f_{j+1}^{n-1}(S_{j+1}, Q_j) \right].$$

(6.4)

Type 3 assumes that the probability of inflow is conditional and current inflows are unknown. Previous inflows are employed as the state variable while deriving the steady-state operation policy by means of initial reservoir storage, previous inflow, and optimal release target.

Type 4

$$f_j^n(S_j) = \underset{R_j}{\text{Maximize}} \sum_{Q_j} P_{Q(j)} \left[\text{EP}_j + f_{j+1}^{n-1}(S_{j+1}) \right]. \quad (6.5)$$

The probabilistic model expressed by Type 4 assumes that the inflow probability is unconditional. In the model, a single state variable with reservoir storage is required without considering inflow.

In real-world situations, Type 1 and Type 2 SDP models need inflow forecasts to determine releases as current inflows are unknown. In contrast, Type 3 and Type 4 SDP models do not require inflow forecasts. The optimal operation policies derived by a SDP model assuming either S_{j+1} or R_j as decision variable will be the same. In general, the use of S_{j+1} is preferred to R_j for Type 1 and Type 2 SDP models due to simple computation. In contrast, R_j instead of S_{j+1} has to be chosen as the decision variable in Type 3 and Type 4 SDP models, even though it causes many calculation difficulties. This is due to the fact that it does not guarantee a constant release throughout the period, or an announcement at the beginning of the period of what the release is to be, if the ending storage S_{j+1} was the decision variable in a Type 3 (or Type 4) SDP model. That is, streamflow forecasts would still be required in determining the precise release in real-time operation. It thus would miss the point in applying the Type 3 (or Type 4) SDP model to avoid the necessity of streamflow forecasts.

6.3.2 Constraints in the model

CONTINUITY

The reservoir storage at the beginning of period $j+1$ can be expressed by the continuity equation:

$$S_{j+1} = S_j + I_j - R_j - E_j; \qquad j = 1, 2, \ldots, N, \qquad (6.6)$$

where

E_j = evaporation and seepage from reservoir during period $j\,(10^6\,\text{m}^3)$,

I_j = inflow during period $j\,(10^6\,\text{m}^3)$,

N = total number of periods,

R_j = release including spillage during period $j\,(10^6\,\text{m}^3)$, and

S_j = initial reservoir storage volume at beginning of period $j\,(10^6\,\text{m}^3)$.

Table 6.1 gives the monthly evaporation depth from the Feitsui Reservoir while the study ignores seepage loss.

STORAGE CONSTRAINT

This constraint ensures that reservoir storage during any period j must be within the limits

$$S_j^{\text{min}} \leq S_j \leq S_j^{\text{max}}; \qquad j = 1, 2, \ldots, N, \qquad (6.7)$$

where

S_j^{min} = dead storage of reservoir $(10^6\,\text{m}^3)$, and

S_j^{max} = total reservoir capacity $(10^6\,\text{m}^3)$.

Table 6.1. *Average monthly evaporation from the Feitsui Reservoir*

Month	Jan.	Feb.	Mar.	Apr.	May	Jun.	Jul.	Aug.	Sep.	Oct.	Nov.	Dec.
Evap. (mm)	13	14	24	33	39	40	59	59	42	27	18	16

Table 6.2. *Maximum and minimum storages of the Feitsui Reservoir*

Years	1987–1996	1997–2013	2014–2030
S_j^{min} ($10^6\,\text{m}^3$)	47.00	35.44	23.37
S_j^{max} ($10^6\,\text{m}^3$)	406.00	385.00	350.00

Table 6.2 gives the values of S_j^{min} and S_j^{max} of the Feitsui Reservoir in different years allowing space for siltation after construction.

WATER DEMAND CONSTRAINT
To satisfy water demand, the water demand constraint can be set as

$$R_j^* \leq R_j \leq R_j^{max}; \qquad j = 1, 2, \ldots, N, \qquad (6.8)$$

where

R_j^{max} = allowable maximum release from joint release capacity of spillways, hydropower, and bottom outlets during period j ($10^6\,\text{m}^3$); 9965.5 m^3/s for Feitsui Reservoir (Sinotech, 1985), and

R_j^* = release demand during period j ($10^6\,\text{m}^3$).

Values of R_j^* depend on the discharge of Nanshih Creek and the incremental inflow downstream of the Feitsui Dam. Based on the historical records, this additional incremental flow (Q_j^*) between the gauges and the diversion point for the drinking water intake can be estimated approximately by the following equation:

$$Q_j^* = 0.1Q_j^{K} + 0.124Q_j^{F}. \qquad (6.9)$$

Hence,

$$
\begin{aligned}
R_j^* &= \text{WD} - Q_j^{K} - Q_j^* \quad &&\text{if } R_j^* > 0 \\
&= 0.0 \quad &&\text{otherwise,}
\end{aligned}
\qquad (6.10)
$$

where

Q_j^{K} = streamflow observed at Kueishan Station (Nanshih Creek) during period j ($10^6\,\text{m}^3$),

Q_j^{F} = streamflow observed at Feitsui Station (Peishih Creek) during period j ($10^6\,\text{m}^3$), and

WD = water demand during period j ($10^6\,\text{m}^3$).

WATER QUALITY CONSTRAINT
This constraint sets a lower limit for the release:

$$R_j^{min} \leq R_j + Q_j^{K} + Q_j^*; \qquad j = 1, 2, \ldots, N, \qquad (6.11)$$

where

R_j^{min} = minimum discharge requirement to meet desired water quality upstream of diversion intake.

HYDRO-ENERGY GENERATION CONSTRAINT
The constraint for firm power generation can also be taken into account in the reservoir operation, if necessary, i.e.,

$$E_j^{min} \leq \text{EP}_j \leq E_j^{max}; \qquad j = 1, 2, \ldots, N, \qquad (6.12)$$

where

E_j^{min} = firm hydro-energy requirement during period j according to contract between Taiwan Power Company and Feitsui Reservoir given in Table 6.3, and

E_j^{max} = allowable maximum hydro-energy production during period j by power plant. Installed capacity of power plant in Feitsui Reservoir is 70 MW.

Energy production at any period t is dependent on the effective head and discharge:

$$\text{EP}_j = 9.81 \times \eta \times Q_j \times H \times t/10^6 \text{ (GWh)}, \qquad (6.13)$$

where

η = overall efficiency of turbine and generator = 0.75,

Q_j = discharge through turbine during period j (m^3/s),

H = effective head (m), and

t = time (h).

6.3.3 Discretization of storage spaces and termination criteria

After having defined the values of S_j^{min} and S_j^{max} given in Table 6.2 and if M discrete values are used to represent the explicit state variable (storage volume), the interval [S_j^{min}, S_j^{max}] may be divided into $M - 1$ equally spaced subintervals. In addition to S_j^{min} and S_j^{max}, values associated with the corner points of those subintervals are selected as the possible discrete values of the storage volume.

Related to the termination of computation in SDP, the steady-state condition is attained if the policies of the same period of two successive annual cycles are identical and the

Table 6.3. *Firm power generation requirement at the Feitsui Reservoir*

Month	Jan.	Feb.	Mar.	Apr.	May	Jun.	Jul.	Aug.	Sep.	Oct.	Nov.	Dec.
Power (GWh)	9.0	8.1	9.0	8.7	9.0	12.4	12.8	12.8	12.4	12.8	8.7	9.0

Source: Huang (1989)

annual increment of the objective function's value between $f_j^n(k, i)$ and $f_j^{n+T}(k, i)$ for all k, i, j are within the allowable limit given in Eq. (6.14).

$$(f_j^{n+T} - f_j^n)/f_j^{n+T} \leq 0.001, \qquad (6.14)$$

where, T denotes the number of stages within the annual cycle.

6.3.4 Transition probability of inflows

Inflows are assumed to have the property of a stationary lag-one Markov chain to account for the hydrological uncertainty. Thus, transition probabilities between subsequent monthly flows are derived and incorporated into the SDP analysis.

Further, the joint transition probability of streamflows is considered concerning the joint probability of the occurrence of discrete streamflow classes at the Feitsui and Kueishan gauging stations. The range of possible average 10-daily inflows at each site, Feitsui or Kueishan gauging station, is divided into only 4 classes individually. That is, 16 joint inflow classes are considered at each time step.

When calculating the recursive equation of SDP, the recursive values do not increase, i.e., the expected benefit or loss value in each cycle decays to zero if all the values are zero in a row of the transition probability matrix. In this study, the dimension of the joint transition probability matrix is 16×16, and only 34 data points (1953–86) are available in each period. The number of possible transitions without observations (i.e., zero-elements in the transitional probability matrix) is excessive, thus it may lead to many zero row vectors. The zero row phenomenon is a severe defect in SDP, and it arises because of limited data and/or too many streamflow classes or both.

He *et al.* (1995) showed that the large number of zero-elements in transition probability matrices was the cause for failing to satisfy the convergence criterion, stabilization of expected annual increment of the objective function value, in the SDP model. They further showed that the substitution of these zeros with reasonably small values was a method to overcome this problem.

6.3.5 Data generation

To retain joint transitional probabilities for streamflows of Nanshih and Peishih Creeks, the problem of zero rows as presented previously has to be eliminated. One way to overcome this difficulty is to generate synthetic streamflow data and use them to derive the transitional probability matrices.

DATA GENERATION WITH CONSTANT PARAMETERS

The lag-one Box–Jenkins standardized model, ARIMA(1, 0, 0), is used to generate synthetic time series as long as needed, until there are no zero-elements in the row vectors of the transition probability matrices, i.e., at least one of the elements in any row vector is nonzero.

Based on the standardized model, ARIMA(1, 0, 0) for example, 100 sequences of the same length as the length of the historical series (34 years) are generated and then the desired statistics of each generated series, $j = 1, 2, \ldots, 100$, are calculated and compared with historical series.

The assumption of a stationary process in parameters like mean, variance, skewness, and correlation coefficient is made in this standardized model. However, the 10-day flows in the Hsintien River have the characteristic of seasonal periodicity. That is, different periods have their particular statistics. In detail, it is possible to generalize the model so that the periodicity in hydrological data is taken into consideration.

DATA GENERATION WITH PERIODIC PARAMETERS

Assuming that for data generation a Box–Jenkins model with periodic coefficients will perform better than that with constant coefficients, the above lag-one Markov model has been modified as a multiseasonal lag-one Markov model. This study considers two types of simulation models; i.e., the lag-one univariate autoregressive model given by Thomas and Fiering (1962) and the lag-one multivariate autoregressive model proposed by Matalas (1967). The former is considered for single sites while the latter concerns not only the autocorrelation but also the cross-correlation between the streamflows observed at the Feitsui and Kueishan gauging stations.

SEASONAL UNIVARIATE AR(1) MODEL (THOMAS–FIERING MODEL)

This model, also known as the Thomas–Fiering model, requires parameter estimation of mean, variance, and lag-one serial correlation for each season. Phien and Ruksasilp

(1981) recommended the use of the Thomas–Fiering model for data generation due to its simplicity and efficiency.

SEASONAL MULTIVARIATE AR(1) MODEL (YOUNG–PISANO MODEL)

Since the joint transition probability of inflows is considered, the simultaneous behavior of cross-correlation between the streamflows of Peishih and Nanshih Creeks, recorded at the Feitsui and Kueishan gauging stations is important. Therefore, the simultaneous simulation of inflows for derivation of the joint transition probabilities may be needed.

After running 100 sequences with length equal to 34 years, the statistics of data generated by the univariate AR(1) model were compared with the historical record. The data generated by the univariate AR(1) model can be accepted with preservation of the mean, standard deviation, skewness, and lag-one serial correlation at each station. However, the univariate AR(1) fails to reproduce the statistics of cross-correlation. On the other hand, the data generated by the multivariate AR(1) model can preserve the mean, standard deviation, skewness, lag-one serial correlation, lag-zero cross-correlation, and lag-one cross-correlation.

The comparison of univariate and multivariate AR(1) models clearly indicates the superiority of the latter over the former in the reproduction of the historical statistics. Thus, in this work multivariate AR(1) is applied to reconstruct the transition probabilities in terms of the 3400-year generated data. In this way the complete zero row vectors have been entirely eliminated.

6.3.6 Comparison among different types of SDP

Out of the four types of SDP models (Eqs. (6.2) through (6.5)), the best one for the Feitsui Reservoir could be found based on test runs with these different models. However, the most suitable one may vary from case to case depending on the circumstances.

Backward-moving stochastic dynamic programming is only applicable in deriving a steady-state operation policy due to the consideration of stochastic inflow (Yeh, 1985). Also, transition probabilities of inflow are assumed to be stationary. A steady-state solution is reached for each SDP model by applying the four different recursive equations up to 36×3 stages.

The expected annual power generation values of the objective function of the Type 3 SDP (404.0 GWh) and the Type 4 models (406.7 GWh) are much higher than those of the other two models, Type 1 and Type 2, whose values are close to each other (306.1 and 308.0 GWh, respectively). For SDP models with more discrete classes in reservoir storage and release, a higher objective function value could be obtained with more

computation time. The Type 3 SDP model needs about five times more computation time than the other three models in deriving the steady-state operation policy, due to consideration of transition probabilities of the inflow for calculation of the system performance value at each stage. Stochasticity is considered for both terms in the right hand side of the recursive equation for the Type 3 SDP model. Operation policies of the Type 1 and Type 2 models are observed to be almost the same. In contrast, the policies of the Type 3 model are close to those of the Type 4 model.

The policies developed by SDP models are sequential ones. These policies depend on each initial storage volume and current inflow for the Type 1 and Type 2 SDP models, on each initial storage volume and the previous inflow for the Type 3 SDP model, and on only initial reservoir storage for the Type 4 model.

The SDP based operation policies are only guidelines. Once developed, the reservoir operation may be simulated and evaluated prior to their actual adoption in practice. Therefore, to study the performances of the derived optimal operation policies based on the four types of SDP, the Feitsui Reservoir system was simulated by using the 34-year (1953–86) historical inflow data. Since the current inflow is not known before the end of the period, perfect inflow forecasting (the one able to predict exactly the observed inflow) is assumed for the Type 1 and Type 2 SDP models during the model simulation due to the requirement of the current-inflow state variable. In contrast, the Type 3 SDP model relies on inflow of the previous period instead of the current one. The inflow requirement is ignored in the Type 4 model.

These simulation results show that the water demand (26.4 m^3/s) could be met each year during the simulation period for all types of SDP models. The average annual generated energy for all SDP models is very similar. The values are 250.19 GWh (Type 1), 250.29 GWh (Type 2), 248.98 GWh (Type 3), and 249.04 GWh (Type 4). The expected values of annual energy generation obtained during the policy development stage are very different from the values obtained from simulations. This is due to consideration of representative values of streamflow during the SDP policy development stage compared to the actual historical streamflows applied in the simulations.

The results indicate that more hydropower would be generated using the Type 1 and Type 2 models assuming perfect forecasts of streamflow are available. However, this is unrealistic since perfect prediction is not possible. In order to select the most suitable SDP model for the Feitsui Reservoir among these four competing models, on-line reservoir operation coupled with observed/forecast streamflow is necessary.

6.3.7 Reservoir operation with Box–Jenkins model based forecast inflows

The aforementioned simulations assumed that the current inflows are known. A model in terms of the current inflow as a state variable would be preferred based on the maximization of hydropower as the objective function. However, in real-world operation the current streamflows at the Feitsui and Kueishan gauging stations are unknown. Consequently, inflow forecasts are needed for the Feitsui reservoir system if a Type 1 or Type 2 SDP model is used. Thus, to put Type 1 and Type 2 SDP models into practice, the linking of the SDP and a forecasting model is required.

In Section 6.3.5, the applicability of the Box–Jenkins model to forecast inflows was discussed. The standardized ARIMA(1, 0, 0) model has been fitted and recommended as an appropriate forecasting model. A more realistic comparison in selecting the most suitable model for the long-term operation of the Feitsui Reservoir can be achieved by applying observed inflow (for Type 3) and forecast inflow by the standardized model (for Type 1 and Type 2) of the Feitsui and Kueishan gauging stations during the period 1985–6. The values of average annual generated energy for the SDP models are 284.78 GWh (Type 1), 291.76 GWh (Type 2), 314.31 GWh (Type 3), and 313.45 GWh (Type 4). Note that the inflow variable is not requested by the Type 4 model.

The Type 3 SDP model, which does not rely on inflow forecasts, shows more hydropower generation and higher water utilization efficiency. The superiority of the Type 3 SDP model over the Type 1 and Type 2 SDP models in the Feitsui reservoir system is based not only on the better operating efficiency obtained, but also on the fact that observed and not forecasted inflow is used. Though the Type 3 model in deriving operation policy needs more computation time compared to others, a powerful computer can tackle it. As compared with the Type 4 model, the Type 3 model performs slightly better.

In order to use the Type 1 (or Type 2) SDP model in real-time operation, streamflow forecasts are needed, and the resulting release will become stochastic due to the use of final storage as the decision variable. The Type 1 SDP model seems better for flood control, in which the storage has to be controlled. In contrast, since the state is known at the beginning of period *j* in the Type 3 (or Type 4) SDP model, the release can be fixed and the final storage becomes a random variable. Previous results show that the Type 1 SDP model performs better than the Type 3 SDP model, if a perfect forecast in streamflows can be made and the release is not stochastic. In reality, with on-line operation, the Type 3 SDP model without consideration of inflow forecasts appears to be better than the other models for water supply and energy

production, where the release is a target. That is, the Type 1 SDP model is preferred if perfect inflow forecasting models are available. Otherwise, the Type 3 SDP model is more appropriate. For the Feitsui reservoir system, due to the inevitable errors existing in forecasting models, the Type 3 SDP model is thus preferred and selected as the most appropriate SDP model for on-line long-term operation.

6.4 OPERATIONAL MODE SWITCH SYSTEM

The operational mode switch (OMS) system developed for determining the reservoir release in a typhoon-prone area is presented next. The OMS model decides whether and when to shift the operation back and forth between long-term "normal mode" and short-term "emergency mode." Use of this model is activated upon a typhoon announcement by the Central Weather Bureau (CWB). The decision criterion to change the operational mode is in terms of the maximization of the expected multiattribute utility trading off flood loss, water shortage, and hydropower production.

The OMS model relies basically on interlinkage of a simulation submodel and a decision submodel as displayed in Figure 6.5. The functions of various parts in the OMS model are described in detail as follows.

6.4.1 Simulation model

The simulation model consists of the elements, (a) data collection, (b) rainfall generation, and (c) runoff simulation, needed to provide the input to the decision model.

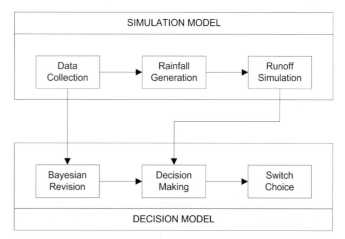

Figure 6.5 Block diagram of operational mode switch

Table 6.4. *Decision making under uncertainty*

	State	Posterior probability	Actions	
			Long-term (1)	Short-term (2)
θ_i	(0–100 mm)	$P(\theta_1|X_j)$	U_{11}	U_{12}
θ_i	(100–200 mm)	$P(\theta_2|X_j)$	U_{21}	U_{22}
θ_i	(200–300 mm)	$P(\theta_3|X_j)$	U_{31}	U_{32}
θ_i	(300–400 mm)	$P(\theta_4|X_j)$	U_{41}	U_{42}
θ_i	(400–600 mm)	$P(\theta_5|X_j)$	U_{51}	U_{52}
θ_i	(600–1000 mm)	$P(\theta_6|X_j)$	U_{61}	U_{62}

DATA COLLECTION

Data collection involves monitoring of incoming data, such as location of the typhoon's eye, wind speed, track and intensity of the approaching typhoon, observed rainfall, and streamflow. Six types of typhoons (X_j) are identified according to their track and intensity.

RAINFALL GENERATION

Rainfall generation produces a probable distribution of the areal rainfall in time for the presumed total amount of precipitation caused by a typhoon. Rainfall generation is based on the transitional probability concept of rainfall depth in consecutive time steps with the assumption of the validity of a first order Markov chain model to describe this process. Transitional probability matrices are distinguished according to different types of typhoons. The generated areal rainfall time series will then become the input to the runoff simulation model to generate the hydrograph by use of a rainfall–runoff model.

RUNOFF SIMULATION

Runoff simulation generates the simulated inflow hydrograph, which serves as a crucial input both to decide the reservoir releases either in normal or in emergency mode, and to determine the operational mode switch. This runoff simulation can be handled through a rainfall–runoff model. Two types of rainfall–runoff models, one based on the unit hydrograph concept and the others of multiple linear regression type, are used.

6.4.2 Decision model

The decision model is interlinked with the simulation model through the outputs of the latter as depicted in Figure 6.5.

BAYESIAN REVISION

Bayesian revision utilizes the Bayes theorem to revise the prior probabilities $P(\theta_i)$ of the possible states of nature (θ_i).

Theoretically, the posterior probabilities $P(\theta_i|X_j)$, the revision of the prior probabilities, are estimated by

$$P(\theta_i|X_j) = P(\theta_i)P(X_j|\theta_i) / \sum_{i=1}^{m} P(\theta_i)P(X_j|\theta_i) \qquad \text{for given} X_j.$$

(6.15)

Based on routinely available meteorological information provided by the Central Weather Bureau, X_j is defined as the observed typhoon type, θ_i (state of nature) on the expected total amount of typhoon-borne rainfall and m on the number of states of nature. Six classes of θ_i are identified (i.e., $m = 6$) ranging from 0 mm to 1000 mm. Furthermore, the likelihood function, $P(X_j|\theta_i)$, of the state of nature is defined based on the on-site historical records.

DECISION MAKING

The decision making element considers the long-term "normal mode" and the short-term "emergency mode" of operations as alternative actions in the switch decision. A decision is made by comparing the multiattribute utilities of the alternative actions. Selection is made following the principle of maximization of the expected multiattribute utility, based on the release policies for the long- or short-term operations over the entire possible state space as shown in Table 6.4. The simulation period depends on the length of the generated rainfall time series.

The utility value denoted by the symbol U with appropriate subscripts designates the pay-offs resulting from each combination of an alternative action and a state of nature. The utility values associated with the possible states of nature and alternative actions can be derived by questioning the real-world decision makers to encompass their preferences for uncertain outcomes.

In assessing the multiattribute utility function (MUF), preferential independence and utility independence are assumed. The former implies that the preference trade-offs are not taken into account among more than two attributes simultaneously, while the latter means that when assessing any particular attribute the decision maker is not influenced by the achievement level of the remaining attributes. For an n-dimensional MUF, $U(\underline{X}) = U(x_1, x_2, \ldots, x_n)$ in which $\underline{X}$ is a vector with n individual attributes, the multiplicative form of MUF under the assumptions of preferential and utility independence satisfies

$$1 + kU(\underline{X}) = \prod_{i=1}^{n}[1 + kk_i U_i(x_i)], \qquad (6.16)$$

where the MUF, $U(\underline{X})$, and the single-attribute utility functions, $U_i(x_i)$, are scaled from 0 to 1. The scaling factors k_i are also within the range between 0 and 1 while the constant k satisfies the following condition:

$$1 + k = \prod_{i=1}^{n}[1 + kk_i]. \qquad (6.17)$$

Having defined the MUF, the problem of switch decision between normal and emergency modes arises. The expected value of the MUF is determined in terms of release from the reservoir.

SWITCH DECISION/CHOICE

The release associated with the normal mode (long-term) is determined by the derived SDP policies according to the synthetic inflow hydrograph. The period of the hydrograph (i.e., simulation period) depends on the duration of generated typhoon-borne rainfall. It has the same duration as the emergency mode (short-term) operation. In real situations, the actual release within a time interval may be different from the SDP based target release. It is assumed here that the SDP policy with the original 10-day basis is also applicable for hourly operation. As previously shown, the steady state release policy derived by the Type 3 SDP model with the inflow of the previous time step as the state variable is appropriate for the Feitsui Reservoir.

In contrast, the most common operational strategy for a regulable flood control reservoir is to keep the outflow at a constant level. Constant release could ensure the complete utilization of the available storage capacity and ensure $S_{max} = S_{cap}$, where S_{max} is the maximum volume of the flood that could be stored in the reservoir, and S_{cap} is the reservoir capacity. The constant release for the emergency mode is then determined by iteration calculation to ensure the fulfilment of $S_{max} = S_{cap}$. However, the strategy requires that the inflowing flood waves are known in advance. Since it is difficult to obtain a completely reliable flood forecasting model, this principle can be used by repeating the forecasting of the total flood wave during the flood event, and

adjusting the releases according to the criterion of $S_{max} = S_{cap}$ for emergency mode operation. This approach leads to variable reservoir releases as an adaptive operation progresses.

The transformation of water release figures with the given state of nature into the MUF yields a series of utilities for each alternative. The expected value over the simulation period depending upon the rainfall duration of a typhoon can be calculated using

$$\bar{U}(\underline{X}|\theta_i) = \xi[U(\underline{X}|\theta_i)] = \sum_{j=1}^{T} U_j(\underline{X}|\theta_i)/T; \qquad \text{for given } \theta_i,$$

$$(6.18)$$

where

$U_j(\underline{X}|\theta_i)$ = multiattribute value at time j for given state of nature (θ_i), and

T = length of simulation period during typhoon attack, equal to length of predicted typhoon-borne flood inflow.

Weighing the expected value of each alternative with the posterior probabilities of the state of nature where the posterior probabilities were just updated from the typhoon observation on the basis of the Bayes theorem, it yields

$$\bar{U}(\underline{X}) = \sum_{i=1}^{m} P(\theta_i|X_j)\,\bar{U}(\underline{X}|\theta_i); \qquad \text{for given } X_j, \qquad (6.19)$$

where

$P(\theta_i|X_j)$ = revision of prior probabilities for given typhoon class (X_j) based on observation from Central Weather Bureau, and

m = number of classes of total typhoon-borne rainfall depth ($m = 6$).

Since the decision criterion of the switch mode is in terms of the maximization of the expected multiattribute utility, the switch action associated with the maximum value of the multiattribute utility between "normal mode" and "emergency mode" is then selected.

The operational switch is updated hourly corresponding to the new observations, either the typhoon condition or the hourly rainfall and discharge observations. That is, if the latest typhoon situation is known, given by the Weather Bureau, revision of the prior probabilities of the state of nature (possible rainfall caused by typhoons) is required. The rainfall hyetograph and the inflow hydrograph are then revised through the "simulation model" within the OMS system. Thereafter, the MUF for each alternative (long-term or short-term operation) has to be reassessed based on the new information to determine the choice for the switch operation.

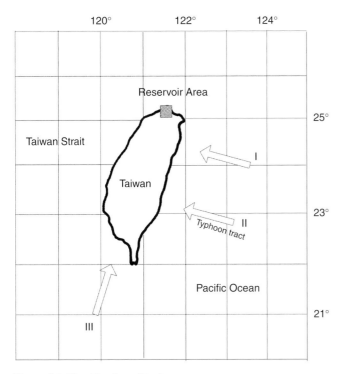

Figure 6.6 Classification of typhoons

Table 6.5. *Summary of information for evaluating multiattribute utility function*

X_i	Worst	Best	$U_i(X_i)$	k_i
X_1	1	0	$1.276 - 0.276\exp(1.531X_1)$	0.700
X_2	1	0	$1.749 - 0.749\exp(0.848X_2)$	0.478
X_3	0	1		0.127
k				-0.669

6.5 APPLICATION AND SENSITIVITY ANALYSIS

The application of the OMS model is shown for the operation of the multipurpose (water supply, hydropower generation, flood alleviation) Feitsui Reservoir. The reservoir capacity is $406 \times 10^6\,\mathrm{m}^3$ with allowable maximum release of $10\,000\,\mathrm{m}^3/\mathrm{s}$. The reservoir can be completely emptied within 2 days through the bottom outlets (sluiceway) and additional tunnel spillway, thus providing the technical option to implement the recommendations of the different operation modes; for example, to pre-empty the reservoir to accommodate a forecast typhoon-borne flood wave.

(a) Typhoons are classified into six types (X_j) which are the combinations of three different typhoon tracks, according to their orientation (westward moving above/below latitude $24°$ N and northward moving) as shown in Figure 6.6 and two different typhoon intensities (remarkable or ordinary). Based on historical records, the state of nature (θ_i) for total rainfall caused by the typhoon is also classified into six groups ranging from 0 mm to 1000 mm. The likelihood function $P(X_j|\theta_i)$ of typhoons is then available from the historical records.

(b) A first order Markov chain related to the transition probabilities of hourly rainfall is assumed. The individual transition probabilities of rainfall associated with different types of typhoon are set up on the basis of historical events. While running the OMS model, the remaining rainfall, after deducting the observed rainfall from the total rainfall, is generated according to the appropriate transition probability. A corresponding inflow hydrograph is then simulated through the multiple linear regression type of rainfall–runoff model.

(c) Concerning the multiattribute utility function, three objectives are considered:

(i) to minimize damage loss due to floods;
(ii) to minimize water shortage as a percentage of demand;
(iii) to maximize profit of power generation.

The single-attribute utility functions have been defined through the use of a questionnaire and interviews with the reservoir managers. Table 6.5 and Figure 6.7 summarize the information obtained from the extensive interview procedures. Notice that $\sum_{i=1}^{3} k_i > 1.0$, which indicates that the choice of a multiplicative form is appropriate (Keeney and Raiffa, 1976).

(d) Determine the release for the normal mode by use of SDP according to the simulated inflow hydrograph while the release for emergency mode is obtained by iteration, determining the value of constant release, which maximizes peak reduction.

Finally, the substitution of water release figures into the MUF yields one series of utility values for each alternative. The expected value over the simulation period can then be calculated.

The expected value of each alternative is weighted by the posterior probability of the state of nature, $P(X_j|\theta_i)$, updated from typhoon observations by means of the Bayes theorem. Thus, the appropriate switch action associated with the maximum value of the MUF can be selected.

The ordinary typhoon Nelson, announced by the Central Weather Bureau at 1400 h on August 21, 1985, was situated at latitude $24.0°$ N and longitude $129.1°$ E, i.e., about 760 km southeast of Taipei, and moved northwestward with 18 km/h

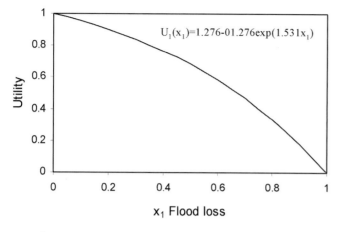

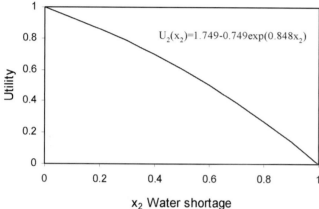

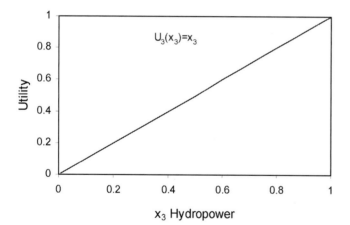

Figure 6.7 Utility functions

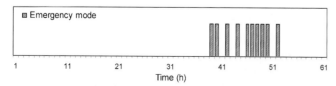

Figure 6.8 Switch process during Typhoon Nelson (August 21–23, 1985)

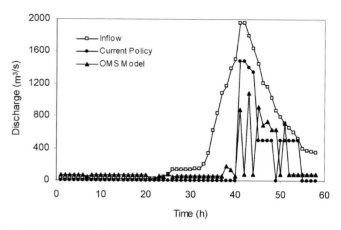

Figure 6.9 Reservoir release during Typhoon Nelson

velocity. Following the aforementioned procedure of the OMS model, in hour 37 (0300 h, August 23, 1985), the operation shifted to emergency mode with 173.42 m³/s release. In the meantime, the operations performed by the OMS model had led to a lower storage level with 342×10^6 m³ to keep sufficient storage for the possible upcoming peak inflow. Figure 6.8 gives the results of the switch decision process during Typhoon Nelson. It shows that the first half-periods still follow the long-term operations. However, the operation changes to emergency mode while the typhoon crosses the catchment of the Feitsui area. The mode returns to normal mode after hour 51 (1700 h, August 23), when the Feitsui area is out of the zone influenced by Nelson at that time.

It can be seen from Figure 6.9 that the peak inflows are concentrated within time intervals 36–47 (0200 h–1300 h) with the amount up to about 2000 m³/s. During these intervals the operational modes are focused on the emergency mode. The peak reduction for the Feitsui Reservoir is not significant within these intervals if the current adopted release policy is applied. For example, at hours 40–43 the inflows approach 2000 m³/s, and the releases are still nearly 1500 m³/s. That is, the peak reduction is only 500 m³/s corresponding to the current policy. In contrast, the efficiency of peak reduction by the OMS model is specified and it is about 1000 m³/s. Figure 6.10 also shows the superiority of operation by the OMS model over that by current policy. Obviously, the operation created by the OMS model tends to lower the water level to maintain sufficient storage before the peak inflows. That is, the "safety margin" operated by the OMS model is larger than that of the current release policy. Furthermore, operation by the OMS model yields 0.9901 units of utilities and 3.6949 GWh, a much better result than the performance by the current policy with 0.9582 utilities and 0.9634 GWh during the simulation period.

Table 6.6. *Sensitivity analysis, the impact of initial storage (Typhoon Nelson, August 21–23, 1985)*

Initial storage ($\times 10^6 \, m^3$)	47	130	230	330	406
MUF by OMS model	0.9538	0.9621	0.9777	0.9901	0.9907
Energy production by OMS model (GWh)	0.0991	0.8772	2.2784	3.6949	3.9926
MUF by present policy	0.9528	0.9532	0.9533	0.9582	0.9723
Energy production by present policy (GWh)	0.0000	0.0365	0.0476	0.9634	2.8645

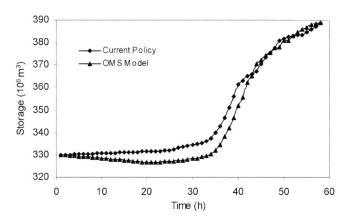

Figure 6.10 Variation of storage during Typhoon Nelson

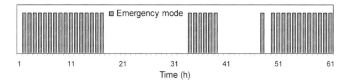

Figure 6.11 Sensitivity analysis of switch with initial storage $406 \times 10^6 \, m^3$ during Typhoon Nelson

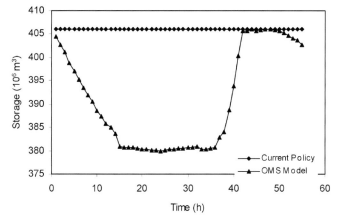

Figure 6.12 Variation of storage with the initial storage of $406 \times 10^6 \, m^3$ during Typhoon Nelson

In fact, the aforementioned OMS operation starts at $330 \times 10^6 \, m^3$. It is expected that operation is quite sensitive to the actual value of the initial reservoir storage while operating the OMS process. For example, if the initial storage is $330 \times 10^6 \, m^3$ and the coming typhoon causes heavy rain, a flood then arises if the operation does not empty the reservoir in advance. On the other hand, the operation results in a water shortage if the reservoir is emptied in advance but the forecasted typhoon does not bring as much rainfall as expected. Thus, caution is required.

It is found that the switch operation of the OMS model is really sensitive to the initial storage. For the case of $47 \times 10^6 \, m^3$, the minimum reservoir storage, there is no occurrence of emergency mode operation during Typhoon Nelson, i.e., the reservoir operations are always kept in the long-term operation mode due to the very low initial water level. However, for the case of $406 \times 10^6 \, m^3$, the maximum reservoir storage, the operation immediately starts in emergency mode to pre-empty the reservoir leaving more storage space for the expected flood inflow caused by Nelson. Figure 6.11 presents the switch process for the typhoon with initial storage $406 \times 10^6 \, m^3$. As Figure 6.12 reveals, the superiority of the OMS model is clearly seen due to the larger safety margin. It has been shown that decisions made with the help of the OMS model are quite reasonable, flexible, and efficient. Table 6.6 also proves the viability of the OMS model. Obviously, the use of the SDP based long-term operation policy, coupled with the OMS supported emergency operation (Duckstein *et al.*, 1989) would provide an improvement over the existing operation rule.

6.6 SOME REMARKS ON OPERATIONAL MODE SWITCH SYSTEM

A methodology for an operational mode switch (OMS) model has been developed for determining the optimal reservoir release under uncertainty during the typhoon season. The OMS model encoding the risk attitude of the decision makers is designed for on-line operation of reservoir systems. It reflects the necessity of short-term operation of reservoir systems facing the occurrence of an extreme event of short duration. The OMS model would provide an efficient tool for decision makers to select the mode and operate reservoir systems during typhoons or under similar conditions.

There are several types of long-term SDP operation policy in terms of different decision and state variables. The analysts are responsible to the decision makers for suggesting a reliable, simple, and understandable policy. The SDP model with initial reservoir storage and past inflow as state variables and release as decision variable (Type 3 SDP model) appears to be the best among all SDP models. However, under different hydrological regimes this "truth" might not be universal and also will depend on the characteristics of the particular water resources system. Considering real forecasts, the Type 3 SDP model seems to be the best for water supply and energy production, where the release is a target. On the other hand, the Type 1 SDP model (initial storage and current inflow as state variables, and ending storage as decision variable) is probably better than the others for flood control where the storage has to be controlled.

The suggested OMS is applicable for the operation of the multipurpose (flood control, water supply, hydropower generation) Feitsui Reservoir in Northern Taiwan. Incorporation of the change of operational mode from a "normal mode" to an "emergency mode" to cope with intermittent typhoon-borne floods enhances the flood control efficiency of the reservoir and the downstream reach without jeopardizing its performance with respect to other purposes. Adoption of the OMS by decision-making authorities, after further analyses of past typhoons and model refinement, is a good possibility. Although uncertainty rules out guaranteeing that the best outcome is obtained, the decisions made by the OMS model are quite reasonable, flexible, and efficient. In case studies of the Feitsui Reservoir, the OMS model generally appears to provide a better policy than the no-switch, long-term policy (SDP). Furthermore, it is simple enough to lead to a rapid transfer of theoretical knowledge into practice.

References

Ampitiya, H. K. (1995). Stochastic dynamic programming based approaches for the operation of a multi-reservoir system. Master's Thesis, Wageningen Agricultural University, the Netherlands.

Ampitiya, H. K., Bogardi, J. J. and Nandalal, K. D. W. (1996). Derivation of optimal operation policies for the reservoirs of the complex Mahaweli water resources scheme in Sri Lanka via a stochastic dynamic programming based approach. In *Proceedings of the International Conference on Aspects of Conflicts in Reservoir Development and Management*, City University, London, UK, pp. 539–548.

Archibald, T. W., McKinnon, K. I. M. and Thomas, L. C. (1997). An aggregate stochastic dynamic programming model of multireservoir systems. *Water Resources Research*, **33**(2), 333–340.

Araujo, A. R. and Terry, L. A. (1974). Operation of a hydrothermal system (in Portuguese), Brazil *Journal of Electrical Engineering*.

Bellman, R. E. (1957). *Dynamic Programming*. Princeton, NJ: Princeton University Press.

Bellman, R. E. and Dreyfus, S. E. (1962). *Applied Dynamic Programming*. Princeton, NJ: Princeton University Press.

Bogardi, J. J., Budhakooncharoen, S., Shrestha, D. L. and Nandalal, K. D. W. (1988). Effect of state space and inflow discretization on stochastic dynamic programming based reservoir operation rules and system performance. In *Proceedings, 6th Congress, Asian and Pacific Regional Division*, IAHR, Vol. 1, Kyoto, Japan, pp. 429–436.

Bogardi, J. J., Brorens, B. A. H. V., Kularathna, M. D. U. P., Milutin, D. and Nandalal, K. D. W. (1995). *Long-term Assessment of a Multi-unit Reservoir System Operation: The ShellDP Program Package Manual*. Report 59, Internal Publication Series, Department of Water Resources, Wageningen Agricultural University, the Netherlands.

Bogardi, J. J. and Milutin, D. (1995). Sequential decomposition in the assessment of long-term operation of large-scale systems. In S. P. Simonović, Z. Kundzewicz, D. Rosbjerg and K. Takeuchi (eds.), *Modeling and Management of Sustainable Basin-Scale Water Resource Systems*, Proceedings of a symposium held at the XXI General Assembly of the International Union of Geodesy and Geophysics (Boulder, 1995). IAHS Publ. No. 231, pp. 233–240.

Bogardi, J. J., Milutin, D., Louati, M. E. and Keser, H. (1994). The performance of a long-term operation policy of multi-unit reservoir systems under drought conditions. In *Proceedings of the VIII IWRA World Congress: Satisfying Future National and Global Demands*, Cairo, Egypt.

Bogardi, J. J. and Szöllösi-Nagy, A. (2004). Towards the water policies for the 21st century: a review after the World Summit on Sustainable Development in Johannesburg. In E. Cabrera and R. Cobacho, (eds.), *Challenges of the New Water Policies for the XXI Century*. Proceedings of the Seminar on Challenges of the New Water Policies for the 21st Century, 2002. Lisse/Abingdon/Exton (PA)/Tokyo: A. A. Balkema Publishers, pp. 17–37.

Brass, C. (2006). Optimising operations of reservoir systems with stochastic dynamic programming (SDP) under consideration of changing objectives and constraints. Ph.D. Dissertation (in German), Ruhr Universitaet Bochum, Germany.

Budhakooncharoen, S. (1986). Comparison of reservoir operation strategies. Master's Thesis No. WA 86–3, Asian Institute of Technology, Bangkok, Thailand.

Budhakooncharoen, S. (1990). Interactive multi-objective decision making in reservoir operation. Doctoral Dissertation, Asian Institute of Technology, Bangkok, Thailand.

Butcher, W. S. (1971). Stochastic dynamic programming for optimum reservoir operation. *Water Resources Bulletin*, **7**(1), 115–123.

Chandramouli, V. and Raman, H. (2001). Multireservoir modeling with dynamic programming and neural networks. *Journal of Water Resources Planning and Management*, **127**(2), 89–98.

Chow, V. T. and Cortes-Rivera, G. (1974). *Applications of DDDP in Water Resources Planning*. Research Report 78, Urbana: University of Illinois, Water Resources Center.

Chow, V. T., Maidment, D. R. and Tauxe, G. W. (1975). Computer time and memory requirements for DP and DDDP in water resource systems analysis. *Water Resources Research*, **11**(5), 621–628.

Cosgrove, W. and Rijsberman, F. (2000). *World Water Vision: Making Water Everybody's Business*. London: Earthscan Publications Ltd.

Crawley, P. D. and Dandy, C. G. (1989). Optimum reservoir operation policies including salinity considerations. In *Hydrology and Water Resources Symposium*, Christchurch, New Zealand, pp. 289–293.

Dandy, C. G. and Crawley, P. D. (1990). Optimization of cost and salinity in reservoir operations. In S. P. Simonovic *et al.* (eds.), *Proceedings of the International Symposium on Water Resources Systems Application*, University of Manitoba, Canada, pp. 452–461.

Dandy, C. G. and Crawley, P. D. (1992). Optimum operation of a multiple reservoir system including salinity effects. *Water Resources Research*, **28**(4), 979–990.

Dantzing, G. B. (1963). *Linear Programming and Extension*. Princeton, NJ: Princeton University Press.

Dias, N. L. C., Pereira, M. V. F. and Kelman, J. (1985). Optimization of flood control and power generation requirements in a multi-purpose reservoir. In *Proceedings of the IFAC Symposium on Planning and Operation of Electric Energy Systems*, Rio de Janeiro, Brazil, pp. 121–124.

Duckstein, L., Bogardi, J. J. and Huang, W. C. (1989). Decision rule for switching the operation mode of a multipurpose reservoir. In D. P. Loucks (ed.), *Proceedings of the Symposium on Systems Analysis for Water Resources Management: Closing the Gap Between Theory and Practice*. IAHS Publication No. 180, Baltimore, pp. 151–161.

Faber, B. A. and Stedinger, J. R (2001). SSDP reservoir models using ensemble streamflow prediction (ESP) forecasts. In D. Phelps and G. Sehlke (eds.), *Proceedings of the World Water and Environmental Resources Congress 2001, Bridging the Gap: Meeting the World's Water and Environmental Resources Challenges*, Orlando, FL, USA.

Fan, K-Y. D., Shoemaker, C. A. and Ruppert, D. (2000). Regression dynamic programming for multiple-reservoir control. In *Proceedings of Building Partnerships – 2000 Joint Conference on Water Resource Engineering and Water Resources Planning & Management* (doi 10.1061/40517(2000)424).

Fan, K-Y. D., Shoemaker, C. A. and Ruppert, D. (2001). Stochastic multiple-reservoir optimization using regression dynamic programming. In *Proceedings of the World Water and Environmental Resources Congress, Bridging the Gap: Meeting the World's Water and Environmental Resources Challenges* (doi 10.1061/40569(2001)158).

Fontane, D. G., Labadie, J. W. and Loftis, B. (1981). Optimal control of reservoir discharge quality through selective withdrawal. *Water Resources Research*, **17**(6), 1594–1604.

Foufoula-Georgiou, E. and Kitanidis, P. K. (1988). Gradient dynamic programming for stochastic optimal control of multidimensional water resources systems. *Water Resources Research*, **24**(8), 1345–1359.

Gablinger, M. and Loucks, D. P. (1970). Markov models for flow regulation. *Journal of Hydraulics Division, ASCE*, **96**(HY1), 165–181.

Gal, S. (1979). Optimal management of a multireservoir water supply system. *Water Resources Research*, **15**(4), 737–749.

Gilbert, K. C. and Shane, R. M. (1982). TVA hydro scheduling model: theoretical aspects. *Journal of Water Resources Planning and Management*, **108**(1), 1–20.

Goulter, I. C. and Tai, F-K. (1985). Practical implications in the use of stochastic dynamic programming for reservoir operation. *Water Resources Bulletin*, **121**(1), 65–74.

Haimes, Y. Y. (1977). *Hierarchical Analyses of Water Resources Systems*. New York: McGraw-Hill.

Haimes, Y. Y. (1982). Modeling of large scale systems in a hierarchical-multiobjective framework. *Studies in Management Science and Systems*, 7: 1–17, Amsterdam: North-Holland Publishing Company.

Hall, W. A. and Buras, N. (1961). The dynamic programming approach to water resources development. *Journal of Geophysical Research*, **66**(2), 517–520.

Hall, W. A. and Dracup, J. A. (1970). *Water Resources Systems Engineering*, New York: McGraw-Hill.

Hall, W. A. and Shepard, R. W. (1967). *Optimum Operation for Planning of a Complex Water Resources System*, Technical Report 122 (UCLA-ENG 67–54), Water Resources Center, School of Engineering and Applied Science, University of California, USA.

Harboe, R. (1987). Application of optimization models to synthetic hydrologic samples. In *Proceedings of the International Symposium on Water for the Future*, Rome, Italy.

Harboe, R., Gautam, T. R. and Onta, P. R. (1995). Conjunctive operation of hydroelectric and thermal power plants. *Water Resources Journal, ST/ESCAP/SER.C/186*: 54–63 (reprint from *Journal of Water Resources Planning and Management*, **120**(6), 1994).

Harboe, R. C., Mobasheri, F. and Yeh, W. W-G. (1970). Optimal policy for reservoir operation. *Journal of Hydraulic Division, ASCE*, **96** (HY11), 2297–2308.

He, Q., Nandalal, K. D. W., Bogardi, J. J. and Milutin, D. (1995). *Application of Stochastic Dynamic Programming Models in Optimization of Reservoir Operations: A Study of Algorithmic Aspects*. Report 56, Internal Publication Series, Department of Water Resources, Wageningen Agricultural University, the Netherlands.

Heidari, M., Chow, V. T., Kokotovic, P. V. and Meridith, D. D. (1971). Discrete differential dynamic programming approach to water resources systems optimizations. *Water Resources Research*, **7**(2), 273–282.

Hillier, F. S. and Lieberman, G. J. (1990). *Introduction to Operations Research*. New York: McGraw-Hill.

Howard, R. A. (1960). *Dynamic Programming and Markov Processes*. Cambridge, MA, USA: MIT Press.

Huang, W-C. (1989). Multiobjective decision making in the on-line operation of a multipurpose reservoir, Doctoral Dissertation, Asian Institute of Technology, Bangkok, Thailand.

Huang, W.-C., Harboe, R. and Bogardi, J. J. (1991). Testing stochastic dynamic programming models conditioned on observed or forecasted inflows. *Journal of Water Resources Planning and Management*, **117**(1), 28–36.

Jacoby, H. D. and Loucks, D. P. (1972). Combined use of optimization and simulation models in river basin planning. *Water Resources Research*, **8**(6), 1401–1414.

Jaworski, N. A., Weber, W. J. and Deininger, R. A. (1970). Optimal reservoir releases for water quality control. *Journal of the Sanitary Engineering Division, ASCE*, **96**(SA3), 727–740.

Johnson, S. A., Stedinger, J. R., Shoemaker, C. A., Li, Y. and Tejada-Guibert, J. A. (1993). Numerical solution of continuous-state dynamic programs using linear and spline interpolation. *Operations Research*, **41**(3), 484–500.

Karamouz, M. and Houck, M. H. (1982). Annual and monthly reservoir operating rules generated by deterministic optimization. *Water Resources Research*, **18**(5), 1337–1344.

Karamouz, M. and Houck, M. H. (1987). Comparison of stochastic and deterministic dynamic programming for reservoir operating rule generation. *Water Resources Bulletin*, **23**(1), 1–9.

Karamouz, M. and Mousavi, S. J. (2003). Uncertainty based operation of large scale reservoir systems: Dez and Karoon experience. *Journal of the American Water Resources Association*, **39**(4), 961–975.

Karamouz, M. and Vasiliadis, H. V. (1992). Bayesian stochastic optimization of reservoir operation using uncertain forecasts. *Water Resources Research*, **28**(5), 1221–1232.

Keeney, R. L. and Raiffa, H. (1976). *Decisions with Multiple Objectives: Preferences and Value Tradeoffs*. New York: John Wiley and Sons.

Kelman, J., Cooper, L. A., Hsu, E. and Yuan, Sun-Quan (1988). The use of probabilistic constraints in reservoir operation policies with sampling stochastic dynamic programming. In *Proceedings of the 3rd Water Resources Operations and Management Workshop*, Colorado, USA, pp. 1–9.

Kelman, J., Stedinger, J. R., Cooper, L. A., Hsu, E. and Yuan, Sun-Quan (1990). Sampling stochastic dynamic programming applied to reservoir operation. *Water Resources Research*, **26**(3), 447–454.

Kitanidis, P. K. and Foufoula-Georgiou, E. (1987). Error analysis of conventional discrete and gradient dynamic programming. *Water Resources Research*, **23**(5), 845–858.

Kularathna, M. D. U. P. (1992). Application of dynamic programming for the analysis of complex water resources systems: a case study on the Mahaweli river basin development in Sri Lanka. Ph.D. Dissertation. Wageningen Agricultural University, the Netherlands.

Kularathna, M. D. U. P. and Bogardi, J. J. (1990). Simplified system configurations for stochastic dynamic programming based optimization of multireservoir systems. Water resources systems application. In S. P. Simonovic *et al.* (eds.), *Proceedings of the International Symposium on Water Resources Systems Application*, University of Manitoba, Canada.

Kumar, D. N. and Baliarsingh, F. (2003). Folded dynamic programming for optimal operation of multireservoir system. *Water Resources Management*, **17**, 337–353.

Laabs, H. and Harboe, R. (1988). Generation of operating rules with stochastic dynamic programming and multiple objectives. *Water Resources Management*, **2**, 221–227.

Labadie, J. W. and Fontane, D. G. (1986). Objective space dynamic programming approach to multidimensional problems in water resources. In A. O. Esogbue (ed.), *Proceedings of the Bellman Continuum-Special NSF Workshop on Dynamic Programming and Water Resources*, Georgia Institute of Technology, Atlanta, USA.

Lane, W. L. and Frevert, D. K. (1989). *Applied Stochastic Techniques, User Manual*. Bureau of Reclamation, Engineering Research Center, Denver, Colorado.

Larson, R. E. (1968). *State Incremental Dynamic Programming*. New York: Elsevier.

Liang, Q., Johnson, L. E. and Yu, Y-S. (1996). A comparison of two methods for multiobjective optimization for reservoir operation. *Water Resources Bulletin*, **32**(2), 333–340.

Loucks, D. P., Stedinger, J. R. and Haith, D. A. (1981). *Water Resources Systems Planning and Analysis*. Englewood Cliffs, NJ: Prentice-Hall.

Loucks, D. P. and van Beek, E. (2005). In *Water Resources Systems Planning and Management: An Introduction to Methods, Models and Applications* (with contributions from J. R. Stedinger, J. P. M. Dijkman and M. T. Villars), Studies and Reports in Hydrology, Paris: UNESCO Publishing.

Maidment, D. R. and Chow, V. T. (1981). Stochastic state variable dynamic programming for reservoir systems analysis. *Water Resources Research*, **17**(6), 1578–1584.

Matalas, N. C. (1967). Mathematic assessment of synthetic hydrology. *Water Resources Research*, **3**(4), 937–945.

Mawer, P. A. and Thorn, D. (1974). Improved dynamic programming procedures and their practical application to water resource systems. *Water Resources Research*, **10**(2), 183–190.

Meier, W. L. and Beightler, C. S. (1967). An optimization method for branching multistage water resources systems, *Water Resources Research*, **3**(9), 645–652.

Millennium Development Goals (MDGs) (2000). UNITED NATIONS Development Programme (UNDP) Website www.undp.org/mdg/basics.shtml, viewed March 10, 2006.

Millennium Ecosystem Assessment (2005). *Ecosystems and Human Well-Being: Current State and Trends. Findings of the Condition and Trends Working Group*. Millennium Ecosystem Assessment Series. Washington, DC: Island Press.

Milutin, D. (1998). Multiunit water resource systems management by decomposition, optimization and emulated evolution. Ph.D. Dissertation, Department of Water Resources, Wageningen Agricultural University, the Netherlands.

Mousavi, S. J., Karamouz, M. and Menhadj, M. B. (2004). Fuzzy-state stochastic dynamic programming for reservoir operation. *Journal of Water Resources Planning and Management*, **130**(6), 460–470.

Murray, D. M. and Yakowitz, S. J. (1979). Constrained differential dynamic programming and its application to multireservoir control. *Water Resources Research*, **15**(5), 1017–1027.

Nandalal, K. D. W. (1986). Operation policies for two multipurpose reservoirs of the Mahaweli Development Scheme in Sri Lanka. M.Eng. Thesis No.WA-86-9, Asian Institute of Technology, Bangkok, Thailand.

Nandalal, K. D. W. (1995). Reservoir management under consideration of stratification and hydraulic phenomena. Ph.D. Dissertation, Department of Water Resources, Wageningen Agricultural University, the Netherlands.

Nandalal, K. D. W. (1998). The use of optimization techniques in planning and management of complex water resources systems. In *Proceedings of the National Conference Status and Future Direction of Water Research in Sri Lanka*, pp. 119–129.

Nandalal, K. D. W. and Ampitiya, H. K. (1997). The assessment of long-term operation of multi-unit reservoir systems. *Engineer, Journal of the Institution of Engineers, Sri Lanka*, **26**(2), 16–24.

Nandalal, K. D. W. and Sakthivadivel, R. (2002). Planning and management of a complex water resources system: case study of Samanalawewa and Udawalawe reservoirs in the Walawe river, Sri Lanka. *Agricultural Water Management*, **57**(3), 207–221.

Nopmongcol, P. and Askew, A. J. (1976). Multi-level incremental dynamic programming. *Water Resources Research*, **12**(6), 1291–1297.

Opricović, S. and Djordjević, B. (1976). Optimal long-term control of a multipurpose reservoir with indirect users. *Water Resources Research*, **12**(6), 1286–1290.

Orlob, G. T. and Simonovic, S. P. (1981). Reservoir operation for water quality control. In *Proceedings of the International Symposium on Real-Time Operation of Hydrosystems*, Waterloo, Ontario, Canada, pp. 599–616.

Palmer, R. N. and Holmes, K. J. (1988). Operational guidance during droughts: expert systems approach. *Journal of Water Resources Planning and Management*, **114**(6), 647–666.

Phien, H. N. and Ruksasilp, W. (1981). A review of single site models for monthly streamflow generation. *Journal of Hydrology*, **52**, 1–12.

Randall, D., Houck, M. H. and Wright, J. R. (1990). Drought management of existing water supply system. *Journal of Water Resources Planning and Management*, **116**(1), 1–20.

Ratnayake, U. R. (1995). Sequential stochastic optimization of a reservoir system. Doctoral Dissertation No. WA 95-2, Asian Institute of Technology, Bangkok, Thailand.

Reznicek, K. K. and Simonovic, S. P. (1990). An improved algorithm for hydropower optimization, *Water Resources Research*, **26**(2), 189–198.

Reznicek, K. K. and Simonovic, S. P. (1992). Issues in hydropower modeling using the GEMSLP algorithm, *Journal of Water Resources Planning and Management*, **118**(1), 54–70.

Roefs, T. G. and Bodin, L. D. (1970). Multireservoir operation studies. *Water Resources Research*, **6**(2), 410–420.

Rogers, D. F., Plante, R. D., Wong, R. T. and Evans, J. R. (1991). Aggregation and disaggregation techniques and methodology in optimization. *Operations Research*, **39**(4), 553–582.

Saad, M. and Turgeon, A. (1988). Application of principal component analysis to long-term reservoir management. *Water Resources Research*, **24**(7), 907–912.

Saad, M., Turgeon, A. and Stedinger, J. R. (1992). Censored-data correlation and principal component dynamic programming. *Water Resources Research*, **28**(8), 2135–2140.

Saad, M., Turgeon, A., Bigras, P. and Duquette, R. (1994). Learning disaggregation technique for the operation of long-term hydroelectric power systems. *Water Resources Research*, **30**(11), 3195–3202.

Shane, R. M. and Gilbert, K. C. (1982). TVA hydro scheduling model: practical aspects. *Journal of Water Resources Planning and Management*, **108**(1), 21–36.

Shiati, K. (1991). Salinity management in river basins: modelling and management of the salt-affected Jarreh Reservoir. Doctoral Dissertation, Wageningen Agricultural University, the Netherlands.

Shrestha, D. L. (1987). Optimal hydropower system configuration considering operational aspects, M.Eng. Thesis, Asian Institute of Technology, Bangkok, Thailand.

Shrestha, D. L., Bogardi, J. J. and Paudyal, G. N. (1990). Evaluating alternative state space discretization in stochastic dynamic programming for reservoir operation studies. In S. P. Simonovic *et al.* (eds.), *Proceedings of the International Conference on Water Resources Systems Application*, University of Manitoba, Canada, pp. 378–387.

Simonovic, S. P. (2000). A shared vision for management of water resources. *Water International*, **25**(1).

Simonovic, S. P. and Orlob, G. T. (1981). Optimization of New Melons Reservoir operation for water quality management. In *Proceedings of International Conference on Environmental Systems Analysis and Management*, International Federation for Information Processing, Rome, Italy.

Simonovic, S. P. and Orlob, G. T. (1984). Risk–reliability programming for optimal water quality control. *Water Resources Research*, **20**(6), 639–646.

Sinotech Engineering Consultants Inc. (1985). *Technical Report of the Operational Rules of Feitsui Reservoir*. Sponsored by the Taiwan Power Company.

Stedinger, J. R., Sule, B. F. and Loucks, D. P. (1984). Stochastic dynamic programming models for reservoir operation optimization. *Water Resources Research*, **20**(11), 1499–1505.

Su, S. Y. and Deininger, R. A. (1974). Modeling the regulation of Lake Superior under uncertainty of future water supplies. *Water Resources Research*, **10**(1), 11–25.

Tai, F. K. and Goulter, I. C. (1987). A stochastic dynamic programming based approach to the operation of a multireservoir system. *Water Resources Bulletin*, **23**(3), 371–377.

Takeuchi, K. (2002). Future of reservoirs and their management criteria. In J. J. Bogardi and Z. W. Kundzewicz (eds.), *Risk, Reliability, Uncertainty, and Robustness of Water Resources Systems*, Cambridge, UK: Cambridge University Press, pp. 190–198.

Teixeira, A. S. and Marino, M. A. (2002). Coupled reservoir operation–irrigation scheduling by dynamic programming. *Journal of Irrigation and Drainage Engineering*, **128**(2), 63–73.

Tejada-Guibert, J. A., Johnson, S. A. and Stedinger, J. R. (1993). Comparison of two approaches for implementing multireservoir operation policies derived using stochastic dynamic programming. *Water Resources Research*, **29**(12), 3969–3980.

Tejada-Guibert, J. A., Johnson, S. A. and Stedinger, J. R. (1995). The value of hydrologic information in stochastic dynamic programming models of a multireservoir system. *Water Resources Research*, **31**(10), 2571–2579.

Thomas, H. A. and Fiering, M. B. (1962). Mathematical synthesis of streamflow sequences for the analysis of river basins by simulation. In A. Mass *et al.* (eds.), *Design of Water Resources Systems*, Cambridge, MA: Harvard University Press, pp. 459–493.

Tilmant, A., Vanclooster, M., Duckstein, L. and Persoons, E. (2002). Comparison of fuzzy and nonfuzzy optimal reservoir operation policies, *Journal of Water Resources Planning and Management*, **128**(6), 390–398.

Trott, W. J. and Yeh, W. W-G. (1973). Optimization of multiple reservoir systems. *Journal of the Hydraulics Division, ASCE*, **99**(HY10), 1865–1884.

Turgeon, A. (1980). Optimal operation of multireservoir power systems with stochastic inflows. *Water Resources Research*, **16**(2), 275–283.

Turgeon, A. (1981). A decomposition method for the long-term scheduling of reservoirs in series. *Water Resources Research*, **17**(6), 1565–1570.

Turgeon, A. (1982). Incremental dynamic programming may yield non-optimal solutions. *Water Resources Research*, **18**(6), 1599–1604.

Umamahesh, N. V. and Chandramouli, S. (2004). Fuzzy dynamic programming model for optimal operation of a multipurpose reservoir. In S. Herath, A. Pathirana and S. B. Weerakoon (eds.), *Proceedings of the International Conference on Sustainable Water Resources Management in the Changing Environment of the Monsoon Region*, Vol. II, November, Sri Lanka, pp. 552–557.

UN (1992). *Agenda 21*. United Nations Conference on Environment & Development, Rio de Janerio, Brazil, 3 to 14 June 1992. www.un.org/esa/sustdev/documents/agenda21/english/Agenda21.pdf.

Vasiliadis, H. V. and Karamouz, M. (1994). Demand-driven operation of reservoirs using uncertainty-based optimal operation policies. *Journal of Water Resources Planning and Management*, **120**(1), 101–114.

Vedula, S. and Kumar, D. N. (1996). An integrated model for optimal reservoir operation for irrigation of multiple crops. *Water Resources Research*, **32**(4), 1101–1108.

Verhaeghe, R. J. and Tholan, N. (1983). *Illustrative Examples of Optimization Techniques for Quantitative and Qualitative Water Management*. Report on Investigation, Report R999-8. Delft Hydraulics Laboratory, the Netherlands.

WCED (World Commission on Environment and Development) (1987). *Our Common Future*. New York: Oxford University Press.

World Commission on Dams (2000). *Dams and Development: A New Framework for Decision-Making. The Report of the World Commission on Dams*. London: Earthscan Publications Ltd.

World Water Assessment Programme (2003). *Water for People Water for Life: The United Nations World Water Development Report*. UNESCO Publishing and Bernan Association.

WSSD (2002). *Johannesburg Plan of Implementation*. World Summit on Sustainable Development, August 26–September 4, 2002, Johannesburg, South Africa. www.un.org/esa/sustdev/documents/WSSD_POI_PD/English/WSSD_PlanImpl.pdf.

Wurbs, R. A. (1993). Reservoir-system simulation and optimization models. *Journal of Water Resources Planning and Management*, **119**(4), 455–472.

Yakowitz, S. (1982). Dynamic programming applications in water resources. *Water Resources Research*, **18**(4), 673–696.

Yeh, W. W-G. (1985). Reservoir management and operation models: a state-of-the-art review. *Water Resources Research*, **21**(12), 1797–1818.

Yekom Consulting Engineers (1980). *Shapur and Dalaki Project Feasibility Report, Jarreh Storage Dam*. Vol.1: Engineering.

Young, G. K. (1967). Finding reservoir operation rules. *Journal of the Hydraulics Division, ASCE*, **93**(HY6), 297–321.

Index